AF588144

Synthesis Lectures on Communications

This series of short books cover a wide array of topics, current issues, and advances in key areas of wireless, optical, and wired communications. The series also focuses on fundamentals and tutorial surveys to enhance an understanding of communication theory and applications for engineers.

Mohammad Abdul Matin •
M. Rezwanul Mahmood

Machine Learning in Next Generation Multiple Access (NGMA)

Mohammad Abdul Matin
Electrical and Computer Engineering
North South University
Dhaka, Bangladesh

M. Rezwanul Mahmood
Electrical and Computer Engineering
North South University
Dhaka, Bangladesh

ISSN 1932-1244 ISSN 1932-1708 (electronic)
Synthesis Lectures on Communications
ISBN 978-3-032-19419-0 ISBN 978-3-032-19420-6 (eBook)
https://doi.org/10.1007/978-3-032-19420-6

This Springer imprint is published by the registered company Springer Nature Switzerland AG
The registered company address is: Gewerbestrasse 11, 6330 Cham, Switzerland

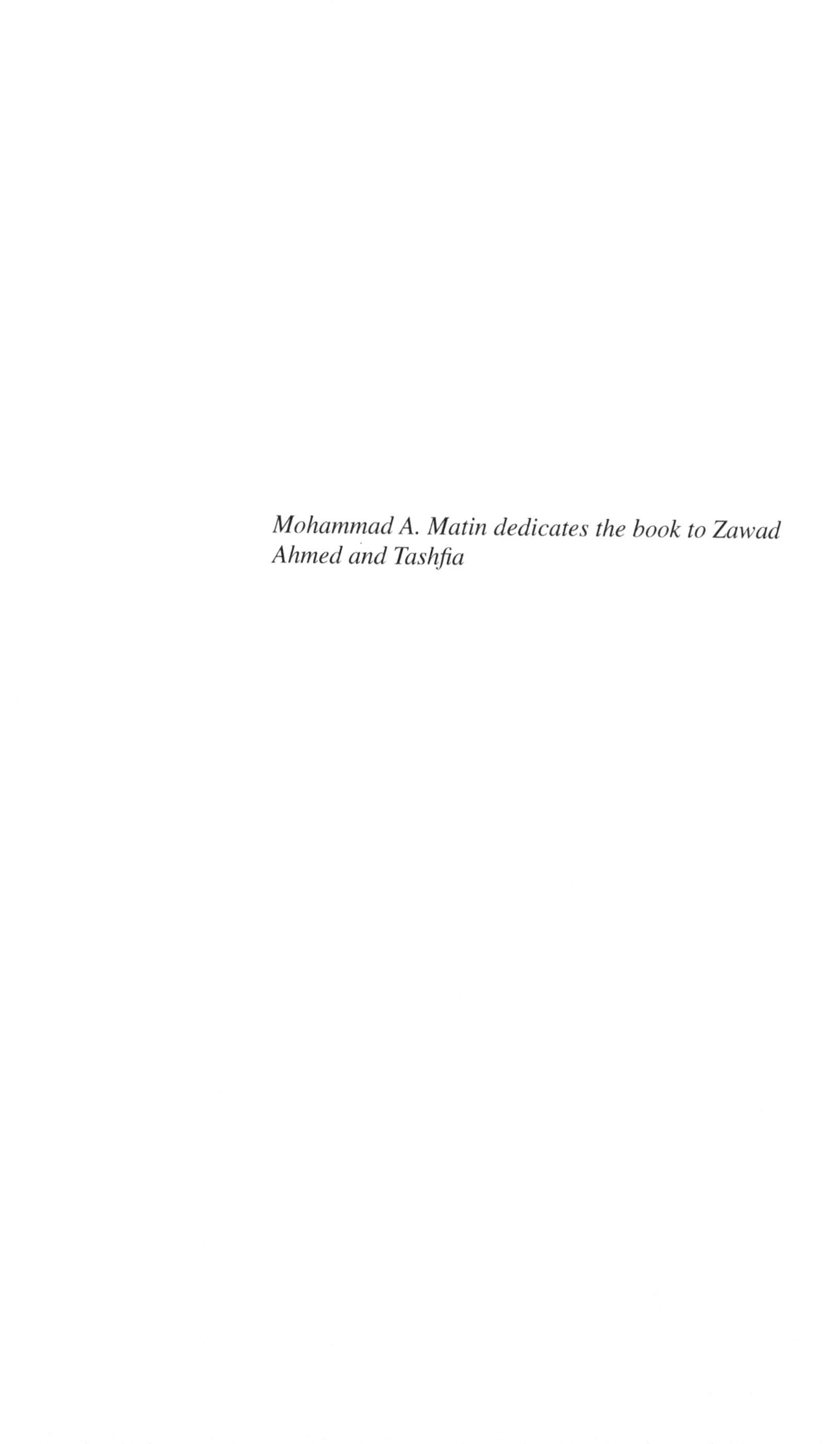

Mohammad A. Matin dedicates the book to Zawad Ahmed and Tashfia

Contents

Acronyms

1G	First Generation
2G	Second Generation
3G	Third Generation
3GPP	Third Generation Partnership Project
4G	Fourth Generation
5G	Fifth Generation
ADMM	Alternating Direction Method of Multipliers
AMPS	Advanced Mobile Phone System
ANN	Artificial Neural Network
AP	Access Point
AR	Augmented Reality
AutoGNN	Automated learning Graph Neural Network
B5G	Beyond 5G
BC	Broadcast Channel
B-FS	Brute-Force Search
BLA	Bayesian Learning Automata
BS	Base Station
BSDC	Best SINR association with Density-based Spatial Clustering of Applications with Noise
CDMA	Code Division Multiple Access
CD-NOMA	Code-Domain NOMA
CNN	Convolutional Neural Network
CPU	Central Processing Unit
CSI	Channel State Information
D4PG	Distributed Distributional Deterministic Policy Gradient
DDPG	Deep Deterministic Policy Gradient
DECT	Digital European Cordless Telephone
DL	Deep Learning
D-NLUPA	Divide and Next Largest difference based User Pairing Algorithm
DNN	Deep Neural Network
DoF	Degree of Freedom

D-OMA	Delta-Orthogonal Multiple Access
DPC	Dirty Paper Coding
DQN	Deep Q-Network
DRL	Deep Reinforcement Learning
DS-CDMA	Direct Sequence CDMA
DT-DPG	Deep Transfer Deterministic Policy Gradient
EE	Energy Efficiency
ELM	Extreme Learning Machine
FAMA	Fluid-Antenna Multiple Access
FDD	Frequency Division Duplexing
FDMA	Frequency Division Multiple Access
GD-Local	Greedy Local
GD-Offload	Greedy Offload
GEVD	Generalized Eigenvalue Decomposition
GSM	Global System for Mobile Communication
GSVD	Generalized Singular Value Decomposition
IC	Interference Channel
IDFT	Inverse Discrete Fourier Transform
IDMA	Interleave Division Multiple Access
IoT	Internet of Things
IS-54	Interim Standard-54
IS-95	Interim Standard-95
ISaC	Integrated Sensing and Communications
LDMA	Location Division Multiple Access
LDPC	low-Density Parity Check
LReLU	Leaky ReLU
LSTM	Long Short-Term Memory
LTE	Long Term Evolution
LTE-A	LTE-Advanced
MA	Multiple Access
MAC	Multiple Access Channel
MAQ-learning	Multi-Agent Q-learning
MDP	Markov Decision Process
MEC	Mobile Edge Computing
ML	Machine Learning
MLP	Multilayer Perceptron
MMSE	Minimum Mean Squared Error
mMTC	Massive Machine-Type Communication
MPA	Message Passing Algorithm
MRC	Maximal-Ratio Combining
MRT	Maximum Ratio Transmission
MSE	Mean Squared Error

NGMA	Next-Generation Multiple Access
NGWN	Next-Generation Wireless Network
NLAF	NonLinear Adaptive Filter
NOMA	Non-Orthogonal Multiple Access
NOMA-CS/CB	NOMA with Coordinated Scheduling/Beamforming
NOMAD	Non-linear Optimization algorithm with the Mesh Adaptive Direct search
NOMA-JP	NOMA with Joint Processing
OFDM	Orthogonal Frequency Division Multiplexing
OFDMA	Orthogonal Frequency Division Multiple Access
OMA	Orthogonal Multiple Access
PDC	Pacific Digital Cellular
PDMA	Pattern Division Multiple Access
PD-NOMA	Power-Domain NOMA
PHS	Personal Handy phone System
PLAF	Partially Linear Adaptive Filter
QoS	Quality of Service
ReLU	Rectified Linear Unit
ResNet	Residual Network
RIS	Reconfigurable Intelligent Surface
RSMA	Rate Splitting Multiple Access
SAQ-learning	Single-Agent Q-learning
SARSA	State-Action-Reward-State-Action
SCA	Successive Convex Approximation
SCMA	Sparse Code Multiple Access
SDMA	Space Division Multiple Access
SIC	Successive Interference Cancellation
SINR	Signal-to-Interference-plus-Noise Ratio
SRPA	Sum-Rate maximization-based Power Allocation
TD3	Twin Delayed DDPG
TDD	Time Division Duplexing
TDMA	Time Division Multiple Access
UC	User Clustering
ULA	Uniform Linear Array
VEC	Vehicular Edge Computing
VLC	Visible Light Communications
VR	Virtual Reality
WCDMA	Wideband CDMA
ZF	Zero-Forcing

Introduction

1

1.1 Introduction

The next-generation wireless communication aims to deliver ultrafast data rates of up to 1 Tb/s, exceptional spectral efficiency of 100 b/s/Hz, ultralow latency below 1 ms, and connectivity for up to 10 million devices/km^2. This ambitious vision supports Internet of Everything (IoE). Based on these specifications and the broader goals of 6G-based next-generation wireless networks, it is evident that such systems must not only meet stringent communication requirements but also connect diverse types of device providing ubiquitous coverage, integrate multiple functionalities, and incorporate native intelligence. Consequently, advanced multiple access (MA) technologies, such as next-generation multiple access (NGMA), are expected to replace traditional, communication-oriented MA schemes. These are network access design methods that allow users to connect to the network in specific manners and utilize services provided by the network systems. Such network access designs involve multiplexing of users in time, frequency, and code resources. The simultaneous provision of network services depends on how the users are allowed to share the network resources. Based on this, multiple access techniques are divided into orthogonal multiple access (OMA) and non-orthogonal multiple access (NOMA). The following sections discuss them and their standardization.

1.1.1 Frequency Division Multiple Access (FDMA)

Frequency division multiple access (FDMA) allocates unique channels to each user upon request (Rappaport 2001). Each user does not share the same frequency spectrum in FDMA. The *Advanced Mobile Phone System* (AMPS), which was the first U.S. analog

M. Abdul Matin, M. R. Mahmood, *Machine Learning in Next Generation Multiple Access (NGMA)*, Synthesis Lectures on Communications,
https://doi.org/10.1007/978-3-032-19420-6_1

cellular system in the first generation (1G), utilized FDMA/frequency division duplexing (FDD). It allowed the user to utilize the allocated channel as two simplex channels, one for the forward channel (used to send voice signals from the base station (BS) to the user k) and the other for the reverse channel (used to send voice signals from the user to BS). The frequency duplexed channels were 45 MHz apart. If B_T, B_C, and B_G represent total bandwidth, channel bandwidth, and guard bandwidth, respectively, then the number of channel N_{FDMA} is computed as follows:

$$N_{FDMA} = \frac{B_T - 2N_B}{B_C}. \tag{1.1}$$

1.1.2 Time Division Multiple Access (TDMA)

In TDMA, several users can access a common frequency spectrum in different time slots within a frame (Rappaport 2001; Falconer et al. 1995). In other words, each user can utilize the available channel bandwidth entirely; however, each of them has to wait for its turn. The TDMA-based network systems conduct data transmission by buffering the desired data in the previous frame and transmitting them in a burst manner in the desired time slot of the next frame. This buffer-and-burst approach for data transmission requires the implementation of digital data and digital modulation in TDMA-based network systems.

A TDMA frame consists of a preamble (containing address and synchronization information utilized by the BS(s) and the user(s)), an information message (containing several time slots, where each of them possesses trail bits, synchronization bits, information data, and guard bits), and tail bits. The guard bits allow the receiver(s) to synchronize with different time slots and frames.

In TDMA with time division duplexing (TDD), the time slots are divided into two parts, one part is for forward link channels, and the other part is for reverse link channels. In TDMA with FDD, the entire time slots are used either for forward or for reverse link channels, but the carrier frequencies for both link channels vary. The second generation (2G) cellular systems such as *Global System for Mobile Communication* (GSM), *Interim Standard-54* (IS-54), and *Pacific Digital Cellular* (PDC) used TDMA with FDD, whereas the *Digital European Cordless Telephone* (DECT) and *Personal Handy phone System* (PHS) used TDMA with TDD (Falconer et al. 1995).

Given the total bandwidth B_T, the channel bandwidth B_C, the guard bandwidth B_G, and the maximum number of TDMA users $K_{(TDMA)}$ supported on each channel, the number of channels N_{TDMA} is computed as follows:

$$N_{TDMA} = \frac{K_{(TDMA)}(B_T - 2N_B)}{B_C}. \tag{1.2}$$

1.1.3 Code Division Multiple Access (CDMA)

In FDMA/TDMA, specific time slots or frequency channels are allocated to each user. In contrast, the Code Division Multiple Access (CDMA) scheme allows communication between the BS(s) and the user(s) through time slot(s)-frequency channel(s) sharing (Rappaport 2001; Shah et al. 2021). However, the message of each user is multiplied with a large user-specific orthogonal bandwidth signal, called a spreading signal or pseudo-codeword. To detect messages transmitted through the CDMA scheme, the receiver(s) require(s) the knowledge of the codeword used by the transmitter. It executes a time correlation operation for desired codeword detection, treating other undesired codewords as noise due to decorrelation.

CDMA is prone to near-far problem scenario, which occurs when many users are allocated to the same channel. In the case of a receiver, e.g., a BS, the strong received signal elevates the noise floor level, creating a possibility for signal transmission failure for the weak signals. To address this problem, the power control mechanism is implemented in CDMA, particularly in direct sequence CDMA (DS-CDMA). This mechanism ensures that the power of the received signals from each user device in both uplink and downlink communications is approximately equal (Rappaport 2001; Zigangirov 2004).

Interim Standard-95 (IS-95) standardized the CDMA-based system in the 2G era, specified for the uplink communication in 824–849 MHz and for the downlink communication in 869–894 MHz. Later, in the third generation (3G) cellular system era, standards such as wideband CDMA (WCDMA) and CDMA2000 are introduced. The earlier standard is formulated on the basis of GSM, while the latter is on the basis of IS-95.

1.1.4 Orthogonal Frequency Division Multiple Access (OFDMA)

Orthogonal frequency division multiple access (OFDMA) utilizes orthogonal frequency division multiplexing (OFDM) in networks with multiple users (Li et al. 2013; Shah et al. 2021). In OFDM, the user's data are divided into multiple parallel streams, where each of them is transmitted over different orthogonal frequencies, known as subcarriers. For instance, the digitally modulated data X are split into N_c sub-streams, where N_c represents the number of subcarriers. After performing a t-point inverse discrete Fourier transform (IDFT) operation on $X[n_c]$ ($n_c \in N_c$), an OFDM symbol is formed, which is given below:

$$x_t = \sum_{n_c=0}^{N_c-1} X[n_c] \exp\left(\frac{j2\pi n_c t}{T}\right), t \in T. \tag{1.3}$$

The orthogonality in OFDMA handles the near-far problem scenario by allocating the users with different subcarrier frequencies. In addition, the flexibility in bandwidth allocation addresses the time-varying traffic requirements and responds to channel fluctuation. Furthermore, the decomposition of a given channel into parallel flat-frequency subcarriers simplifies MIMO signal processing. It also allows the exploitation of frequency selectivity of the channel and interference.

The Long-Term Evolution (LTE), a project of the telecommunication association named 3rd Generation Partnership Project (3GPP), standardizes the OFDMA for the fourth generation (4G) era, with bandwidths ranging within 1.4–20 MHz (Li et al. 2013). The LTE-Advanced (LTE-A) uses up to 100 MHz of bandwidth in several frequency bands such as 450–470 MHz, 698–862 MHz, 790–862 MHz, 2.3–2.4 GHz, 3.4–4.2 GHz, and 4.4–4.99 GHz (Akyildiz et al. 2010).

1.1.5 Non-orthogonal Multiple Access (NOMA)

Non-orthogonal multiple access (NOMA) allocates the same time-frequency resources to multiple users. NOMA is primarily categorized into power-domain NOMA (PD-NOMA) and code-domain NOMA (CD-NOMA). The former NOMA category differentiates users in terms of power allocation, and the latter category differentiates them in terms of spreading sequences. The NOMA receiver utilizes successive interference cancellation (SIC) (in case of PD-NOMA) or message passing algorithm (MPA) or sphere decoder-based methods (in case of CD-NOMA) to extract users' signals (Dai et al. 2018; Shah et al. 2021; Liu and Yang 2021).

DCM, Huawei, CATT, Qualcomm, and Nokia have several NOMA-based schemes such as PD-NOMA, sparse code multiple access (SCMA), pattern division multiple access (PDMA), rate-splitting multiple access (RSMA), interleave division multiple access (IDMA), etc. LTE Release 13 and the digital TV standard have included NOMA as MA scheme (Meredith 2015; Zhang et al. 2016). According to NTT DOCOMO, NOMA is recognized as a key multiple access scheme for the fifth generation (5G) network system (Benjebbovu et al. 2013). For beyond 5G (B5G)/sixth generation (6G) network systems, the standardization of NOMA/NOMA-inspired access system such as delta-orthogonal multiple access (D-OMA) (Shah et al. 2021) is still ongoing.

1.2 Driving Force for NOMA

NOMA overcomes the limitations of network access provision for multiple users by allowing them to share a given time-frequency resources, as mentioned in Sect. 1.1.5. This is enabled through the superposition of signals from multiple users in the same time-frequency resources. Consequently, NOMA plays a crucial role in supporting massive connectivity (Shin et al. 2017; Vaezi et al. 2019).

Moreover, the flexibility of network resource allocation in NOMA based on the use cases of user devices and their quality of service (QoS) requirements facilitates low latency communication. In addition, the optimal utilization of available network resources, particularly the frequency spectrum, contributes to improved user fairness and high spectral efficiency. This advantage arises because NOMA users can access a larger fraction, or even the entirety, of the available bandwidth compared to orthogonal multiple access (OMA) users, for whom the allocated frequency spectrum is inversely proportional to the number of users being served.

The key features of NOMA that support the development of NGMA-based wireless networks are discussed in this section:

1. *Massive connectivity*: NOMA utilizes power/code resources to provide network accessibility to several users in the same time-frequency resources. This feature allows NOMA to efficiently utilize given network resource blocks, compared to OMA schemes. Thus, NOMA addresses the challenge of accommodation of numerous Internet of Things (IoT) devices within a given network resource, which is a design criterion of NGMA (Ahmed et al. 2024; Belmekki and Alouini 2024).
2. *High spectral efficiency with user fairness*: By superposing different user signals at allocated power/code, NOMA achieves higher spectrum efficiency than OMA. This also allows higher data transmission within a given spectrum bandwidth while ensuring user-fairness (Vaezi et al. 2019). The criterion of achieving high throughput through NGMA is possible to meet through NOMA. Figure 1.1 presents the achievable regions

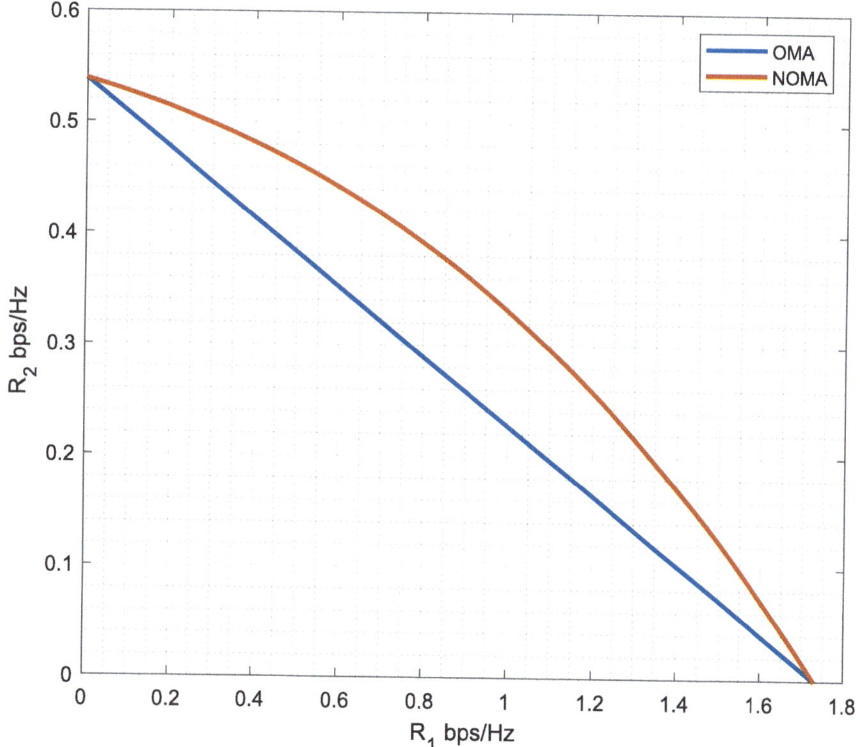

Fig. 1.1 Achievable regions for the two-user OMA and NOMA in a downlink network system. Here, $\gamma_1 = 3\gamma_2 = \sqrt{10}$, and BS transmission power is considered to be 30 dBm

for downlink network systems incorporating NOMA and OMA, which demonstrates the enhanced rate region achievement through NOMA.

3. *Low latency*: Since NOMA allows transmission of several user data over the same frequency and same time, the waiting time of the user for the availability of the channel gets reduced, which results in minimization of latency. Furthermore, incorporation of NOMA with network access request methods such as grant-free and semi-grant-free access methods minimizes the time for scheduling and granting access (Belmekki and Alouini 2024). Therefore, NOMA is an excellent candidate of NGMA in terms of fulfilling the low latency requirement.
4. *Flexibility*: NOMA provides enhanced scalability and flexibility by fine-tuning power allocations (in PD-NOMA), on the basis of channel conditions and QoS requirements. It is beneficial in the network system containing devices with both low and highly constrained QoS requirements. It is also possible by combining NOMA with grant-based and grant-free access schemes. The requirement of the NGMA schemes to adapt to the dynamically changing network conditions and services along with meeting stringent QoS requirements can be realized by NOMA-based schemes.

1.3 Classification of NOMA Schemes

NOMA has been broadly categorized into two categories, one is power-domain NOMA (PD-NOMA), and the other is code-domain NOMA (CD-NOMA). PD-NOMA-based transmission relies on the utilization of time/frequency resources with different power allocation. On the other hand, CD-NOMA-based transmission (inspired from CDMA) relies on the utilization of time/frequency resources with a different spreading sequence. The following discusses these two NOMA schemes.

1.3.1 Power-Domain NOMA (PD-NOMA)

A downlink NOMA network system is considered, where user a and user b receive signals from a BS. Let the signals for user a and user b be denoted by x_a and x_b respectively. The superposed signal is then represented by $x = \sqrt{P\alpha_a}x_a + \sqrt{P\alpha_b}x_b$, where P is the total transmission power, and α_a and α_b are the power allocation coefficients for user a and user b, respectively, such that $\alpha_a + \alpha_b = 1$. The user $k \in [m, n]$ receives the following signal:

$$y_k = h_k x + n_k, \tag{1.4}$$

where h_k and n_k are the channel coefficients and noise, respectively.

Followed by signal detection, the receiver depends on the SIC to extract its own signal from the superposed signal. The user with better channel condition first detects the signal of the user with poor channel condition, subtracts it from the detected superposed signal, and thus obtains its desired signal by decoding the remaining part. The user with poor CSI obtains its own signal directly decoding the superposed signal.

1.3.2 Code-Domain NOMA (CD-NOMA)

Suppose the BS transmits $x = \beta_a c_a s_a + \beta_b c_b s_b$ to user a and user b, where $beta_a$ (β_b), c_a (c_b), and s_a (s_b) are the weights corresponding to the user's distance from the BS, the specified spreading sequence, and the desired signal for user a (user b), respectively. Given the CDMA system is supported by O_C chips, the received signal by user $k \in [m, n]$ in the $o_C \in O_C$th chip is given by

$$y_{o_C,k} = h_{o_C,k} x + n_{o_C,k}, \tag{1.5}$$

where $h_{o_C,k}$ and $n_{o_C,k}$ are the channel and noise between the BS and the user on chip o_C.

If MPA-based method is implemented at the receiver end, the detection is characterized by the factor graph representation that returns the marginal distribution at the observation nodes for the variable node inputs. If a sphere decoder-based method is implemented, the potential solution space for $\hat{x}$ detection is obtained by defining a hyperspherical region with the help of $h_{o_C,k}$ and $y_{o_C,k}$.

1.4 NOMA: From Theory to Implementation

The knowledge of the capacity region of single cell-NOMA and the highest achievable region in a multi-cell NOMA is based on the analysis of the exact and the approximated degree of freedom (DoF) (Vaezi and Vincent Poor 2019). Although techniques such as SIC and dirty paper coding (DPC) (Costa 1983) help achieve better rates compared to OMA, their implementation on the user device can be complex. Apart from that, issues such as imperfect channel state information (CSI), power allocation and clustering in multi-user scenario, etc. require considerations to implement NOMA-based schemes.

1. *Power allocation and clustering*: The management of system throughput and user-fairness depends on network resource allocations to the users. Considering several users in a single group can cause high SIC error, which would negatively impact the system performance. By grouping them into multiple clusters on the basis of factors such as each user's CSI and a number of users to be multiplexed, the SIC operation is simplified and the error is minimized (Islam et al. 2016). Besides, power allocation schemes should consider the availability of CSI, QoS, and power constraints. In addition, the trade-offs between the complexity and performance achievement of the user clustering and power allocation schemes should be considered (Vaezi et al. 2019; Ahmed et al. 2024).

2. *Implementation of SIC*: The effect of imperfect and similar CSIs, decoding order, error propagation while decoding a specific user's signal, etc. are crucial considerations for SIC implementation (Ahmed et al. 2024). Imperfect instantaneous CSI-based SIC causes decoding errors, which consequently affects the performance of NOMA-based networks. In case of similar CSIs of the users, NOMA does not provide significant network performance gain compared to OMA, rather it complicates the network system. Therefore, advanced SIC solutions are required to address the abovementioned issues.
3. *Beamforming designs*: Beamformers in NOMA aim at minimizing the inter-user/inter-cluster interference and intercell interference. Along with SIC decoder design, beamforming design plays an important role in providing satisfactory system performance. In terms of network security, the decoding ability of the users in the NOMA network raises privacy concerns. Cryptography and traditional key agreement algorithm-based approaches do not minimize security and privacy risks in sustainable manner. Thus, physical layer security, commonly based on beamforming-based methods, is considered as a catalyst along with cryptography/key agreement algorithm approach to ensure network security (Pakravan et al. 2023).

1.5 Conclusion

This chapter provides an overview of the multiple access techniques in the evolution of the wireless network systems. It discusses NOMA, its theoretical background, and how it can achieve higher capacity compared to OMA. It highlights the role of NOMA in developing NGMA systems. This chapter ends with an overview of the considerations for the practical implementation of NOMA.

References

Ahmed A, Xingfu W, Hawbani A, Yuan W, Tabassum H, Liu Y, Qaisar MUF, Ding Z, Al-Dhahir N, Nallanathan A, et al. (2024) Unveiling the potential of NOMA: A journey to next generation multiple access. IEEE Communications Surveys & Tutorials

Akyildiz IF, Gutierrez-Estevez DM, Reyes EC (2010) The evolution to 4G cellular systems: LTE-Advanced. Physical communication 3(4):217–244

Belmekki BEY, Alouini MS (2024) NOMA as the next-generation multiple access in nonterrestrial networks. Proceedings of the IEEE

Benjebbovu A, Li A, Saito Y, Kishiyama Y, Harada A, Nakamura T (2013) System-level performance of downlink NOMA for future LTE enhancements. In: 2013 IEEE GLOBECOM workshops (GC Wkshps), IEEE, pp 66–70

Costa M (1983) Writing on dirty paper (corresp.). IEEE transactions on information theory 29(3):439–441

Dai L, Wang B, Ding Z, Wang Z, Chen S, Hanzo L (2018) A survey of non-orthogonal multiple access for 5G. IEEE Communications Surveys & Tutorials 20(3):2294–2323

Falconer DD, Adachi F, Gudmundson B (1995) Time division multiple access methods for wireless personal communications. IEEE Communications Magazine 33(1):50–57

Islam SR, Avazov N, Dobre OA, Kwak KS (2016) Power-domain non-orthogonal multiple access (NOMA) in 5G systems: Potentials and challenges. IEEE communications surveys & tutorials 19(2):721–742

Li J, Wu X, Laroia R (2013) OFDMA mobile broadband communications: A systems approach. Cambridge University Press

Liu Z, Yang LL (2021) Sparse or dense: A comparative study of code-domain NOMA systems. IEEE Transactions on Wireless Communications 20(8):4768–4780

Meredith JM (2015) Study on downlink multiuser superposition transmission for LTE. In: TSG RAN Meeting, vol 67

Pakravan S, Chouinard JY, Li X, Zeng M, Hao W, Pham QV, Dobre OA (2023) Physical layer security for NOMA systems: Requirements, issues, and recommendations. IEEE Internet of Things Journal 10(24):21721–21737

Rappaport TS (2001) Wireless communications: principles and practice. Cambridge University Press

Shah AS, Qasim AN, Karabulut MA, Ilhan H, Islam MB (2021) Survey and performance evaluation of multiple access schemes for next-generation wireless communication systems. IEEE Access 9:113428–113442

Shin W, Vaezi M, Lee B, Love DJ, Lee J, Poor HV (2017) Non-orthogonal multiple access in multi-cell networks: Theory, performance, and practical challenges. IEEE Communications Magazine 55(10):176–183

Vaezi M, Vincent Poor H (2019) NOMA: An Information-Theoretic Perspective, Springer International Publishing, Cham, pp 167–193

Vaezi M, Schober R, Ding Z, Poor HV (2019) Non-orthogonal multiple access: Common myths and critical questions. IEEE Wireless Communications 26(5):174–180

Zhang L, Li W, Wu Y, Wang X, Park SI, Kim HM, Lee JY, Angueira P, Montalban J (2016) Layered-division-multiplexing: Theory and practice. IEEE Transactions on Broadcasting 62(1):216–232

Zigangirov KS (2004) Theory of code division multiple access communication. John Wiley & Sons

Machine Learning (ML) in NOMA: Toward NGMA Networks

2

2.1 Introduction

Wireless network operations like resource allocation, user clustering, channel state information (CSI) estimation, and signal detection require low-complexity algorithmic methods to improve network performance. Traditional methods often fail to maintain the expected trade-off between accurate network performance and computational complexity. Besides, these methods often require information that is difficult to acquire. For instance, it can be difficult to accurately determine traffic loads in advance, which poses a challenge for implementing the baseline scheme for BS switching as outlined in Li et al. (2014). To address such issues, machine learning (ML) techniques are implemented to execute several network operations by learning the features of the provided data.

Any ML-based methodology typically uses a neural network architecture to create an approximated model of any continuous function. From the perspective of wireless network systems, such method/technique aims in developing a simple mathematical model that can replicate any complicated mathematical formulation of a given network system. Appropriate learning mechanisms and learning functions play crucial roles in designing effective ML methods to perform several network operations.

2.2 Machine Learning Techniques

This section presents the following ML learning techniques which are supervised, unsupervised, reinforcement, federated, and transfer learning.

M. Abdul Matin, M. R. Mahmood, *Machine Learning in Next Generation Multiple Access (NGMA)*, Synthesis Lectures on Communications,
https://doi.org/10.1007/978-3-032-19420-6_2

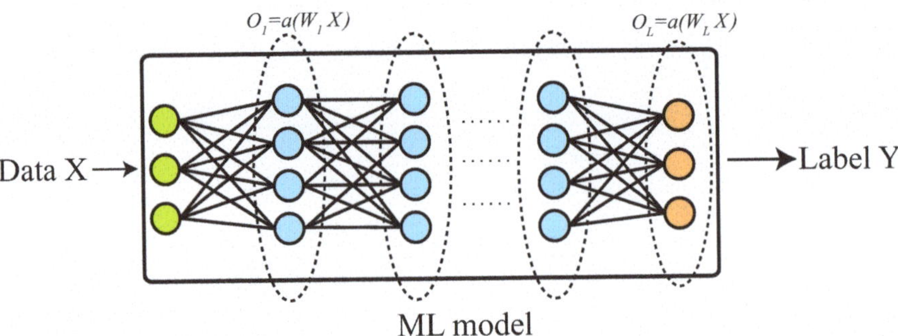

Fig. 2.1 Supervised learning

2.2.1 Supervised Learning

Supervised learning techniques are used to determine the relationship between the given data and their respective labels. Figure 2.1 provides a demonstration of the supervised learning techniques. For instance, $T = (X, Y)$ represents a dataset where X is the data and Y is the label corresponding to the data. In this case, supervised learning determines the mapping between X and Y through training. Let us split the dataset $T = (X, Y)$ into training dataset $T_{train} = (X_{train}, Y_{train})$ and testing dataset $T_{test} = (X_{test}, Y_{test})$. The former dataset assists in creating the supervised learning-based ML model, and the latter dataset assists in validating the obtained output.

Considering a single hidden layer-based supervised learning incorporated ML model and following (2.14), we can write the following equation:

$$O_L = a(W_L X + b_L). \tag{2.1}$$

Here, O_L is the output, and W_L is the learning parameter Lth layer of the ML model. Initially, the model provides errors when compared with the label Y in the training stage, whose mean squared error (MSE) is given by

$$MSE_{sup} = \frac{\sum_{j=1}^{J}(O_{j,train} - Y_{j,train})^2}{J}. \tag{2.2}$$

Here, J is the data size. To minimize the training error, parameters W_L and b_L are updated iteratively. Stochastic gradient descent and Adam optimizers are commonly used optimizers to update these parameters.

2.2.2 Unsupervised Learning

Unsupervised learning techniques are used to identify hidden patterns in a given unlabeled dataset. Figure 2.2 provides a demonstration of the unsupervised learning techniques. A

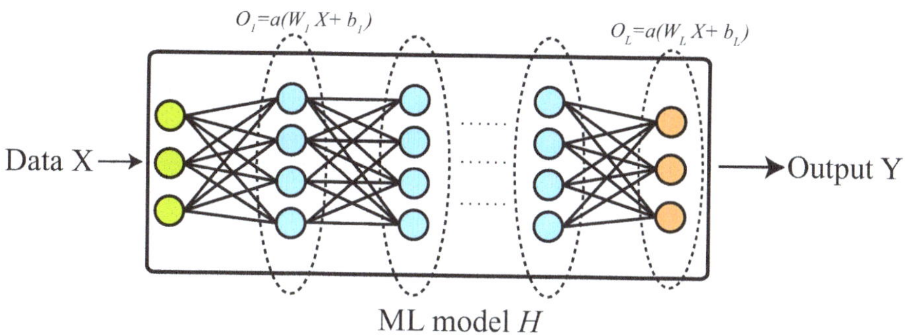

Fig. 2.2 Unsupervised learning

loss function is utilized to train an unsupervised learning-based ML model. For example, X is a given dataset, and Y is the outcome of the learning model H. The mathematical expression of the hidden layer output of the ML model is the same as (2.1). To update the model parameters, this learning model will use an objective function, called loss function $\mathcal{L}_M$. Therefore, the loss function is given by

$$\min_{y_j \in Y} \mathcal{L}_M(x_j, H; y_j). \tag{2.3}$$

It is to be noted that, similar to supervised learning techniques, unsupervised learning techniques also do not initially provide the expected output by utilizing (2.3). Therefore, optimizers such as stochastic gradient descent and Adam are required to update the ML model parameters.

2.2.3 Reinforcement Learning

Reinforcement learning techniques involve learning about a given environment by an agent, analyzing the state of the environment, and taking action to maximize a reward. Figure 2.3 provides a demonstration of the reinforcement learning techniques. Markov decision process (MDP) acts as a mathematical framework for these types of learning techniques (Mahmood et al. 2022).

Let us define a state space $s_1, s_2, s_3, \ldots, s_t \in \mathcal{S}$ and action space $a_1, a_2, a_3, \ldots, a_t \in \mathcal{A}$. The agent will be guided by a policy, which is expressed as

$$\pi(a|s) = \mathbb{P}[a_t = a|s_t = s]. \tag{2.4}$$

To determine the best course of action for an agent to take, reinforcement learning techniques use a state-/action-value function for optimal policy calculation. The following equation represents the state-value and action-value expressions:

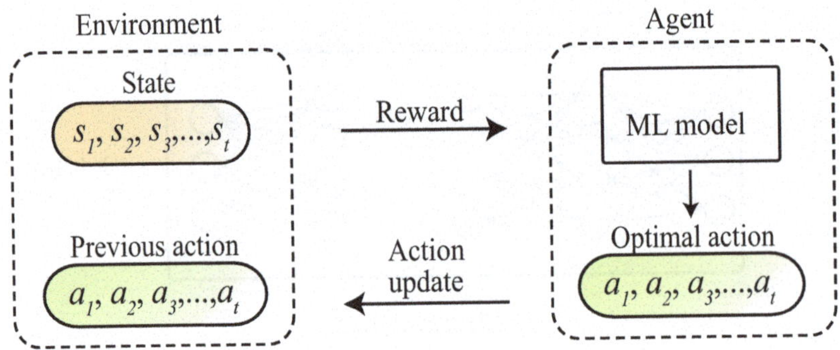

Fig. 2.3 Reinforcement learning

$$v_\pi(s) = \mathbb{E}_\pi[G_t|s_t = s] \tag{2.5a}$$

$$q_\pi(s,a) = \mathbb{E}_\pi[G_t|s_t = s, a_t = a]. \tag{2.5b}$$

Here, G_t represents the reward function at time t. Generally, there are three types of reward models, namely finite horizon ($\mathbb{E}\left[\sum_{t=0}^{T} g_t\right]$), discounted-infinite horizon ($\mathbb{E}\left[\sum_{t=0}^{\infty} \gamma^t g_t\right]$), and average reward ($\lim_{t\to\infty} \mathbb{E}\left[\frac{1}{t}\sum_{t=0}^{t} g_t\right]$) (Van Otterlo and Wiering 2012). If discounted, the infinite horizon reward model is considered. Then (2.5a) can be defined in terms of *Bellman equation*, which is given by

$$v_\pi(s) = \mathbb{E}_\pi\left[\sum_{t=0}^{\infty} \gamma^t g_t|s_t = s\right] \tag{2.6a}$$

$$= \sum_{s'} T(s, \pi(a|s), s')\left(G(s, a, s' + \gamma v_\pi(s'))\right). \tag{2.6b}$$

Here, T is the transition function. It is to be noted that v_π is unique for any given policy π. For an optimal policy (expressed as π^*), $v_\pi^*(s) \geq v_\pi(s)$, $\forall s \in \mathcal{S}$, and $\forall \pi$. If $v^* = v_\pi^*$, we can write the following equation:

$$v^*(s) = \max_{a\in\mathcal{A}} \sum_{s'\in\mathcal{S}} T(s, a, s')\left(G(s, a, s' + \gamma v_\pi(s'))\right). \tag{2.7}$$

The optimal action under the state-value function given in (2.7) is chosen with the help of the following rule:

$$\pi^*(s) = \arg\max_a \sum_{s'\in\mathcal{S}} T(s, a, s')\left(G(s, a, s' + \gamma v_\pi(s'))\right) \tag{2.8a}$$

$$= \arg\max_a q^*(s, a), \tag{2.8b}$$

where

$$q^*(s,a) = \sum_{s' \in S} T(s,a,s') \left(G(s,a,s' + \gamma \max_{a'} q^*(s',a')) \right). \tag{2.9}$$

The relation between the state-value and the action-value function for optimal policy is given by

$$v^*(s) = \max_a q^*(s,a). \tag{2.10}$$

Reinforcement learning techniques are used in the scenario where the accurate distribution of T and reward are not known (Van Otterlo and Wiering 2012). In that case, the action-value function is first learned, and later, the state-value function is derived. Here, the reward acts as a guide during the determination of the optimal policy.

2.2.4 Federated Learning

Distributed ML techniques use several local computing devices, located at different locations, to develop a global ML model through parallel computation (Khan et al. 2021). These techniques mitigate the feasibility issue of the centralized ML techniques in the case of large data size and the locations of the source of such a high volume of data. However, the computational capability of the local computing devices (in terms of cycles/sec, storage size), the communication resources, and the data heterogeneity (in terms of nonindependent and identical distribution of data) make the implementation of distributed learning techniques challenging. Furthermore, data-parallel-based (where the training data are distributed among the local computing devices to train a given ML model), or model-parallel-based (where the same data are utilized by the local computing devices to train different segments of a given ML model) distributed ML models utilize data sent from multiple network devices. Therefore, these techniques do not guarantee data privacy. Federated learning techniques, instead of requiring the transfer of data from local devices, involve the transfer of local ML parameters learned by these computing devices to the central computing device, which builds up a global model. Figure 2.4 provides a diagrammatic representation of federated learning.

Let $\mathbb{L}$ be the number of devices, each containing $k_{\mathbb{L}} \in d_{\mathbb{L}}$ data points. The objective function of the federated learning techniques is provided below (2.11):

$$\min_{\boldsymbol{w}_{M_L}} \frac{1}{K_{\mathbb{L}}} \sum_{\mathbb{L}} \sum_{k=1}^{k_{\mathbb{L}}} f_k(\boldsymbol{w}_{M_L}). \tag{2.11}$$

Here, $\boldsymbol{w}_{M_L}$ and $K_{\mathbb{L}}$ represent the global weight parameter(s) and total data points (contained by all the local devices for training their own ML model M_L), respectively.

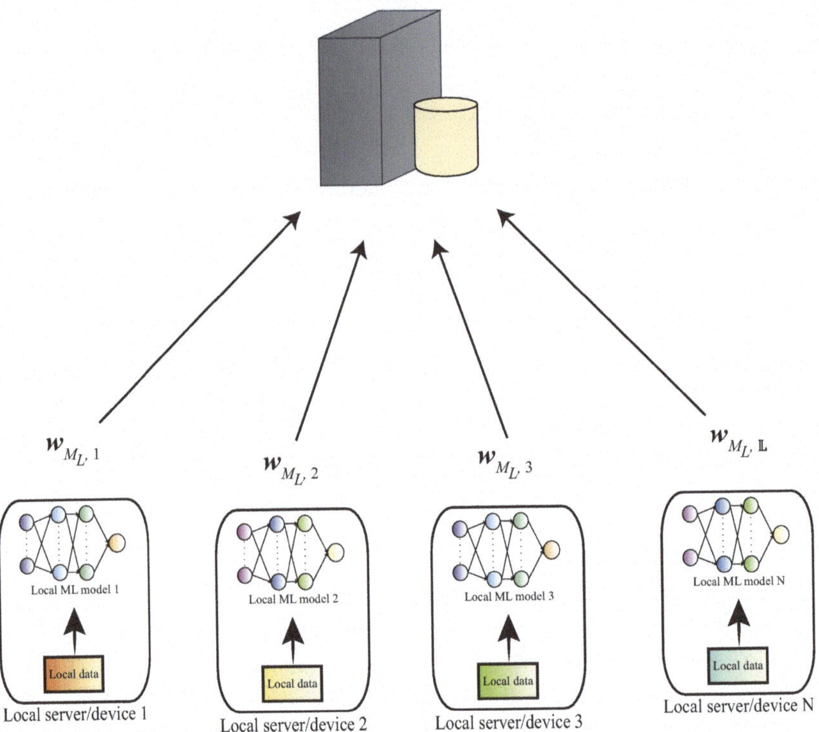

Fig. 2.4 Federated learning

Before updating the global ML model, the local devices first split their own data $d_{\mathbb{L}}$ into batches (say, $\mathcal{B}$). Later, for each batch ($b \in \mathcal{B}$), it uses the stochastic gradient descent technique (or a similar technique) to update their local weight parameter(s) $w_{M_L,\mathbb{L}}$.

$$\boldsymbol{w}^{update}_{M_L,\mathbb{L}} \leftarrow \boldsymbol{w}_{M_L,\mathbb{L}} - \eta \nabla M_L(\boldsymbol{w}_{M_L,\mathbb{L}}, b). \tag{2.12}$$

Here, η and $\nabla M_L(w_{\mathbb{L}}, b)$ represent learning rate and average gradient on batch b at $\mathcal{M}_{\mathbb{L}}$ before update period. The operation (2.12) occurs iteratively at local devices.

At the global server end, for each global iteration, $\mathbb{L}_s \in \mathbb{L}$ clients are randomly selected, where each of these clients computes $\nabla \frac{1}{K_{\mathbb{L}_s}} \sum_{k=1}^{k_{\mathbb{L}_s}} f_k(\boldsymbol{w}_{M_L})$. After aggregating the local models, the global weight update is carried out as follows:

$$\boldsymbol{w}^{update}_{M_L} \leftarrow \boldsymbol{w}_{M_L,\mathbb{L}} - \eta \sum_{\mathbb{L}_s} \frac{k_{\mathbb{L}_s}}{K_{\mathbb{L}}} \left(\nabla \frac{1}{K_{\mathbb{L}_s}} \sum_{k=1}^{k_{\mathbb{L}_s}} f_k(\boldsymbol{w}_{M_L}) \right). \tag{2.13}$$

Federated learning techniques do not fully eliminate the possibility of data leakage. Therefore, the sharing of encrypted local ML models and implementation of data security-

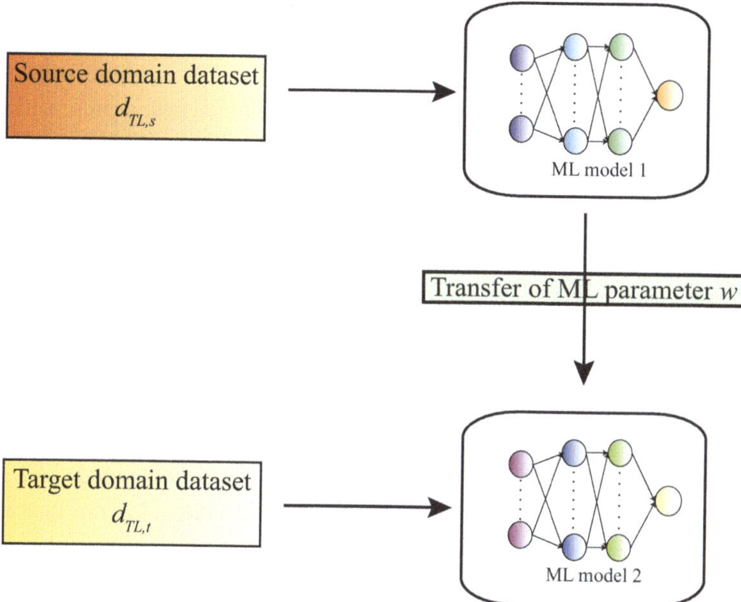

Fig. 2.5 Transfer learning

concerned aggregation algorithm (which considers aggregation of local ML models without decrypting them) must be considered (Brik et al. 2020).

2.2.5 Transfer Learning

In general, training datasets and testing datasets are derived from the same statistical distribution. However, this is not always the case in real-world scenarios. Apart from this, lack of data and insufficient computation capability are common problems faced by traditional ML techniques. Transfer learning techniques aim at addressing these issues (Yuan et al. 2020; Niu et al. 2021). Figure 2.5 presents the diagrammatic representation of transfer learning. These techniques contain two domains, namely, the source domain and the target domain. The source domain performs the main ML training. The target domain refines the trained ML model at the source domain and performs the desired task. This training load distribution addresses the issue of lack of labeled data, as well as high computation time and resource requirements. Furthermore, by combining the knowledge learned by ML in different domains, transfer learning techniques improve the performance of the incorporated ML techniques, whose degradation would have been caused by data distribution mismatch.

Several transfer learning techniques are surveyed in Pan and Yang (2010). In this subsection, the transfer learning technique used for optimization of adapting beamforming for the balancing problem of SINR is presented (Yuan et al. 2020). Let the source domain and the target domain be represented by TL_s and TL_t, respectively. The dataset in the source domain and the target domain is denoted by $d_{TL,s}$ and $d_{TL,t}$, respectively. In the source domain, $d_{TL,s}$ is utilized to tune the ML parameter(s), represented by w, to minimize the loss function given by $w \leftarrow w - \eta \nabla L_{TL_s}(w)$, where η and $\nabla L_{TL_s}(w)$ are learning rate and gradient of the loss function, respectively. In the target domain, $d_{TL,t}$ is utilized to fine-tune w (fully or partially), to minimize $L_{TL_s}(w)$ by using $w \leftarrow w - \eta \nabla L_{TL_t}(w)$ or by using Adam optimizer.

2.3 Activation Functions in ML for NOMA

2.3.1 Introduction to Activation Functions

Let d_t represent data fed to the input of an ML model with a single hidden layer. In this case, the output of the hidden layer neuron o, observed at the output layer, is then expressed as

$$o = f(h) = a(w_h d_t + b_h). \tag{2.14}$$

In case of the ML model with more than one hidden layer (l=[1,2,. . .,L]), Eq. (2.14) can be rewritten as follows:

$$i = f(h_l) = a(w_{h_l} i_{l-1} + b_{h_l}). \tag{2.15}$$

Here, i_{l-1} is the input from $(l-1)$th hidden layer, and $i = f(h_l)$ is the output observed at the lth hidden layer. In both cases, $w_h(w_{h_l})$ and $b_h(b_{h_l})$ denote the weight and bias assigned between the input and the hidden layer, respectively. Here, $a(.)$ is the activation function, which can be a linear function or any nonlinear function such as sigmoid, hyperbolic tangent function, rectified linear unit function, etc. These activation functions help in designing ML neural network-based approximation functions for data interpretation (Parhi and Nowak 2020). The mathematical expressions of activation functions are described below.

Linear Function

For a given variable x, the linear function gives output ranging from $-\infty$ to $+\infty$. The following equation represents the linear function:

$$f(x) = x. \tag{2.16}$$

Sigmoid Function

If x is a given variable, the sigmoid function will provide the following output:

$$f(x) = \frac{1}{1 + \exp(-x)}. \tag{2.17}$$

This function provides output within the range 0-1 and provides an "S"-shaped graph.

Hyperbolic Tangent

The hyperbolic tangent function, represented by tanh, is similar to the sigmoid in terms of the shape of the graph. The output ranges between -1 and 1. The following equation represents the tanh function:

$$f(x) = \tanh(x) = \frac{2}{1 + \exp(-2x)} - 1 \tag{2.18a}$$

$$= \frac{\exp(2x) - 1}{\exp(2x) + 1}. \tag{2.18b}$$

ReLU and Leaky ReLU

ReLU stands for "Rectified Linear Unit," which gives positive output for the positive values of x (the upper limit is ∞) and 0 for the negative values of x. The output of the ReLU function is given by

$$f(x) = \max(0, x). \tag{2.19}$$

Different from ReLU, leaky ReLU (LReLU) provides output for negative values of x. The following mathematical expression represents LReLU:

$$f(x) = \begin{cases} x, & \text{if } x > 0 \\ 0.01x, & \text{otherwise} \end{cases} \tag{2.20}$$

Softmax

The softmax activation function is used to transform given data into probabilities. The probabilities of each value are proportional to the relative scale of each value in the data. The following mathematical expression reflects the softmax function:

$$f(x)_j = \frac{\exp(x_j)}{\sum_{j=1}^{J} \exp(x_j)}. \tag{2.21}$$

Here, $j = (1, 2, \ldots, J)$ represents each entry of the variable x.

2.3.2 Role of Activation Functions in ML-Based NOMA

The choice of activation function(s) in the neural network architecture of the ML-based methodologies depends on the nature of the desired output provided by ML-neural network. For instance, in Yu et al. (2022) the authors use tanh and linear activation functions for user activity detection and channel estimation for grant-free NOMA systems. The reason behind these choices lies in the ability of the tanh and linear functions to perform nonlinear combination in forward layers and expand the range of the values in the output layer, respectively. Softmax function is also used in Kim et al. (2020) for active user detection in grant-free NOMA to provide the probabilistic value of the users currently active. Sigmoid function is used by the neural network architecture proposed in Lin et al. (2021) for signal demodulation in NOMA-VLC system. In case of power allocation coefficients (which lies between 0 and 1), the sigmoid function provides any values between 0 and 1, whereas the summation of the softmax output gives 1, which makes them ideal choices as the activation function for neural networks. For power allocation values (not coefficients) not limited within 0 and 1, ReLU function can also be chosen, which is done in Ali et al. (2021). The use of ReLU and linear activation functions is also observed in Kumaresan et al. (2020), Rajasekaran and Yanikomeroglu (2023) for effective user clustering in NOMA networks. The LReLU is used in the hidden and output layers of the proposed ML architecture in Li et al. (2020) for task offloading operations, which processes the given CSI to deliver time scheduling and task partitioning as outputs.

2.4 Application of ML in NOMA-Based Networks

The discussion on ML applications in NOMA networks includes user clustering, resource allocation, computational task offloading, successive interference cancellation (SIC), and beamforming.

2.4.1 User Clustering

In a NOMA network system, multiple users are assigned the same time/frequency/code resources, leading to a high probability of interference that negatively impacts the overall system rate. To mitigate this, users are categorized into distinct groups that use orthogonal resources (Ding et al. 2015b). Within a group, users share the same resources. Generally, user grouping in a given network is performed based on transmission power, channel gains, and decoding capabilities constraints.

In Chen et al. (2020), Kumaresan et al. (2020, 2021a,b), Rajasekaran and Yanikomeroglu (2023), the authors have adopted supervised learning techniques for user clustering. In Kumaresan et al. (2020, 2021a), Rajasekaran and Yanikomeroglu (2023), the authors have used artificial neural network (ANN) and deep neural network (DNN) for user clustering, whereas in Kumaresan et al. (2021b), the authors have used extreme learning machine (ELM), which accelerates learning duration. Chen et al. (2020) propose generalized user grouping, allowing the users to be part of more than one group and meet desired user power requirements. The authors aim at maximizing throughput (Eq. (2.22)) and effective user grouping through formulation of joint power control and user grouping optimization problem.

$$R_{throughput} = BR_{NOMA} = B\sum_{k=1}^{K}\log_2(1+\gamma_{NOMA,k}). \tag{2.22}$$

Here, B is the system bandwidth. Table 2.1 demonstrates the use of ML-based techniques for user clustering in NOMA networks.

2.4.2 Resource Allocation

Here, resource allocation refers to the allocation of time, frequency, power, code, etc., to users according to a definite multiple access scheme. For NOMA-based network systems, the allocation of power and channels/subchannels plays a key role in exploiting the benefits of NOMA. Several optimal/heuristic techniques are adopted by the researchers to assign power and channels/subchannels optimally, which are considered to be an NP-hard problem (Lei et al. 2015; Liu and Dai 2013). However, the solution space of such an optimization problem is large. Therefore, mathematical techniques that determine nonlinear relationships among the data are required to find the optimal solution. Machine learning techniques use nonlinear functions to discover these relationships and thus are found useful for resource allocation in NOMA networks.

Supervised, unsupervised, reinforcement, federated, and transfer learning techniques are utilized for this purpose. In Yang et al. (2019), the authors have proposed a DNN-based power allocation scheme on the basis of channel gain information. Later, based on the DNN-based power assignment, the user subchannel assignment technique is executed. In Sun et al. (2019), the authors have presented a supervised learning technique, incorporating deep learning (DL) architecture, for optimal decoding order and power allocation. The queue state of the users, channel gains between the users and the transmitter (in this case, satellite), bandwidth, average and peak power, number of time slots, and duration of each time slot are fed into the proposed DL architecture. The output of the DL contains the optimal decoding order and power allocated to each user. Ali et al. (2021) have proposed a DNN architecture, trained also in a supervised manner, for power allocation at source devices and decode-and-forward relay nodes. The channel gains, the power budget of each user in the network system, and the available power at the relay are fed into the DNN architecture, which provides per-user transmit power as output. In Zhang et al. (2020), the authors have studied a semi-supervised learning technique for DL-

Table 2.1 Application of ML for user clustering

References	Application domain	Objective	ML model(s)	ML technique(s)	Dataset	Compared against
(Kumaresan et al. 2020)	User clustering in downlink NOMA	Throughput performance maximization	ANN	Supervised learning	Transmission power, channel gain, and cluster-related information	B-FS-based clustering, random pairing, and D-NLUPA
(Rajasekaran and Yanikomeroglu 2023)	User clustering	Sum-rate (in Mbps/Hz) maximization	ANN	Supervised learning	Each user's instantaneous CSI and SIC decoding capabilities	Comparable performance against NOMA-MEC and NOMA-best beam
(Kumaresan et al. 2021a)	User clustering and power allocation in downlink NOMA	Throughput performance	DNN	Supervised learning	Transmission power, channel gain, and the optimal cluster formation	B-FS-based UC, ANN-UC, dynamic UC, and OMA
(Kumaresan et al. 2021b)	User clustering in downlink NOMA		Extreme learning machine	Supervised learning	Channel gains, initial transmit powers and cluster-related information	ANN-UC and dynamic UC
(Chen et al. 2020)	Joint power control and user grouping in uplink NOMA system	Throughput maximization	Random forest and support vector machine	Supervised learning	Path loss and user grouping labels	Traditional non-overlapping user grouping with power control (Ding et al. 2015b) and OMA

based resource management, such as subchannel and power allocation, in NOMA. For subchannel allocation, some labeled and unlabeled datasets are considered. The labeled dataset contains channel gain and the corresponding subchannel allocation strategy; the unlabeled dataset contains channel gain only. For power allocation, the unsupervised learning technique is adopted. The proposed methodology aims to maximize the energy efficiency of the network system (defined in (2.23) Yang et al. 2021) under QoS, interference, and power limitation constraints.

$$\eta_{EE} = R/P = \frac{BR_{NOMA}}{P} = \frac{B\sum_{k=1}^{K}\log_2(1+\gamma_{NOMA,k})}{P} \tag{2.23}$$

In Zhu et al. (2021), the authors have adopted a reinforcement learning-based algorithm such as deep deterministic policy gradient (DDPG) for optimal power allocation in MIMO-NOMA vehicular edge computing (VEC) networks. The state space of the DDPG includes buffer length, SINR, and position of each vehicular user; the action space includes transmission power and task processing power for each user; the reward space includes power consumption and delay for task completion by each user. Xiao et al. (2017), Zhai et al. (2021) have proposed Q-learning-based power allocation methodologies. In Xiao et al. (2017), the state, action, and reward space contain user SINR, BS power allocation scheme, and utility of the BS averaged over all the previous experiences, respectively. In Ahsan et al. (2021), deep reinforcement learning (DRL) and state-action-reward-state-action (SARSA) are deployed for time/frequency resource block allocation for the uplink NOMA-IoT networks. The proposed scheme(s) consider(s) the number of BSs, subchannels, users in the state space, and execute(s) resource allocation for IoT users. The reward value depends on the sum-rate and the total users associated with BSs and subchannels. The resource allocation scheme in He et al. (2019) includes the power allocation method (based on Zhu et al. 2017) and the channel allocation method (based on attention-based neural network incorporated DRL). The state, action, and reward space include channel information, assigned channel, and users' data rate (after allocating power to the users), respectively. Table 2.2 demonstrates the use of ML-based techniques for resource allocation in NOMA networks.

2.4.3 Task Offloading

Edge computing promotes highly efficient computational resources and power utilization for use cases such as virtual and augmented reality, holographic communication, security monitoring, etc. (Wang et al. 2020; Naouri et al. 2021). Due to the use of the same communication resources offered by NOMA, simultaneous offloading of the edge devices' computational task to the servers through uplink transmission protocol and downloading of the computational results from the servers through downlink transmission protocol becomes possible (Liu et al. 2022). Traditional methods such as game theory, optimization

Table 2.2 Application of ML for resource allocation

References	Application domain	Objective	ML model(s)	ML technique(s)	Dataset	Compared against
(Yang et al. 2019)	Power allocation and user scheduling (in terms of subchannels) for downlink NOMA	Total sum-rate and computational efficiency of the network system	DNN	Supervised learning	Channel gains as input and power allocation as output	Interior point method, maximum power allocation, random power allocation
(Sun et al. 2019)	Long-term power allocation for downlink satellite-IoT NOMA	Achieve the optimal decoding order and power allocation	DL	Supervised learning	Queue, channel state, SIC decoding order, and power allocation	FuS algorithm
(Ali et al. 2021)	Joint power loading solution at source and relaying nodes in decode and forward relay-aided NOMA	System throughput maximization subject to the battery and SIC constraints	DNN	Supervised learning	Channel gains and user uplink power budget as input; per-user transmit power as output	Lagrange-based optimal solution
(Zhang et al. 2020)	User association, subchannel, and power allocation in downlink NOMA	EE maximization under the constraints of QoS, interference, and power limitations	Semi-supervised learning model (for subchannel allocation); DNN (for power allocation)	Semi-supervised (both supervised and unsupervised); DNN (unsupervised)	Semi-supervised learning: (labeled and unlabeled subchannels); DNN (power allocation related data)	Gradient iteration algorithm, equal power allocation, and random power scheme

(Zhu et al. 2021)	Power allocation optimization for MIMO-NOMA-VEC (uplink transmission scenario)	Low power consumption and latency	DDPG	Reinforcement learning	Buffer length, SINR, and vehicle position in state space; Local execution and offloading powers in action space; power consumption and delay in reward space	GD-Local execution first and GD-Offload first policy
(Xiao et al. 2017)	Power allocation of a BS in a downlink NOMA system	Improvement of the system throughput (under user data rate constraint)	Fast Q-learning	Reinforcement learning	User SINR in state space; BS power allocation strategy in action space; utility of the BS averaged over all the previous real experiences in reward space	OMA Q-learning, NOMA Q-learning, NOMA Hotbooting Q-learning, and NOMA Dyna Q-learning
(Zhai et al. 2021)	Power allocation for multi-antenna downlink NOMA networks	Sum-rate maximization subject to user power allocations, total power budget, and each user's QoS	Q-learning	Reinforcement learning	Distances between users and BS in state space; power allocated to users in action space; each user's data rate, system sum-rate (in Mbps), and total transmission power in reward space	NOMA with random power allocation algorithm, OFDMA, and SRPA method (Liu et al. 2019)
(Ahsan et al. 2021)	Resource allocation	Maximization of the average performance of sum-rates	DRL and SARSA	Reinforcement learning	Number of BSs, subchannels, users in state space; resource block allocation in action space; reward function consisting of sum-rate and total users associated with base stations and subchannels	A memoryless method showing maximum achievable rate by exhaustively searching all possible combinations of 3D associations

(continued)

Table 2.2 (continued)

References	Application domain	Objective	ML model(s)	ML technique(s)	Dataset	Compared against
(He et al. 2019)	Channel and power allocation for downlink NOMA	Maximization of system performance (e.g., sum-rate and minimal rate)	Attention-based Neural Network incorporated DRL (mainly for channel allocation)	Reinforcement learning	Channel information in state space, assigned channel in action space; user data rate as reward (after power and channel allocation)	Joint resource allocation method proposed in Zhu et al. (2017) and the exhaustive search technique
(Zhong et al. 2022a)	Joint optimization of the mobile RIS deployment and their phase shifts, and the power allocation policy for users	Optimization of the sum-rate of the users	Federated learning enabled DDPG	Federated learning, reinforcement learning	RIS and user locations, path loss for each user, and channel information in state space; RIS deployment, RISs' phase shifts, and users' power allocation policy in action space; sum-rate in reward space	OMA, fixed RIS locations and no RIS, and without federated learning techniques
(Zhong et al. 2022b)	Path and power allocation managements of intelligent robots in NOMA-based robotic communications	Optimization of the mission conducting efficiency and communication reliability (in terms of sum-rate)	Deep transfer deterministic policy gradient (DT-DPG)	Transfer learning, reinforcement learning	Instantaneous location of the intelligent robot and channel gains in state space; their motion and power allocations in action space; motion and sum-rate-based reward functions in reward space	OMA and traditional DDPG techniques

theory, heuristic algorithm, dynamic programming, etc. are adopted for task offloading. However, under dynamic task computing, transmission traffic, and time-varying channels, the traditional task offloading methodologies face issues in delivering quick and optimum performance. Supervised and reinforcement learning-based methodologies are used for optimal task offloading under these situations.

In Li et al. (2020), the authors propose DL-based techniques such as cyclic branch network and residual network (ResNet) for a NOMA-assisted mobile edge computing (MEC) system, which prioritizes minimization of total energy consumption. Their methodology receives channel state information of the user-BS link and provides task partitioning and time scheduling as the solution. Qian et al. (2020) present a DRL-based algorithm for computation offloading in a NOMA-multi-access-MEC system. The state space considers channel realization between an IoT device and several computing servers, the action space assigns optimal transmission duration, and the reward space calculates the total energy consumed by the IoT device. In Lin et al. (2023), the authors investigate a deep Q-network (DQN) for task offloading decisions and subchannel assignment in a NOMA-MEC network system. Based on the data to be processed, waiting time in a queue, and channel gain, the DQN performs task offloading and subchannel assignment operations to maximize the task satisfaction ratio of long-term arrival delay-sensitive tasks. In Yang et al. (2020), the authors study the execution of computing resource allocation and task offloading decisions, respectively, by the single-agent Q-learning and the multi-agent Q-learning-based methodologies based on the initial state of energy consumption of the MEC servers for task computation. The proposed scheme aims to minimize the total energy consumption for task completion. Table 2.3 demonstrates the use of ML-based techniques for task offloading in NOMA networks.

2.4.4 Successive Interference Cancellation (SIC) and Beamforming

For multiple antenna-based network systems, such as MIMO, both SIC and beamforming techniques are used to decode the received signal (Liu et al. 2022; Higuchi and Benjebbour 2015). In such cases, the SIC decoding and beamforming performance combinedly provide a decent SINR. ML methodologies are explored in NOMA for SIC and beamforming designs and are discussed below.

In Xu et al. (2023), the authors present an unsupervised learning-based technique for SIC and beamforming for multi-cell NOMA networks, by utilizing data channels and interference channels. This technique aims at improving the system sum-rate and reducing information overhead. Sim et al. (2020) proposed a convolutional neural network (CNN)-based SIC scheme. It uses a supervised learning technique for training, which takes the received superimposed signal as CNN input and provides a decoded user signal as CNN output. It aims at mitigating the decoding loss in imperfect SIC scenarios. Awan et al. (2018) also explore a supervised learning technique for the design of a detector in the MIMO-NOMA system to address the issue where high cluster size enhances the spectrum

Table 2.3 Application of ML for task offloading

References	Application domain	Objective	ML model(s)	ML techniques	Dataset	Compared against
(Li et al. 2020)	To obtain a suboptimal time scheduling and task allocation solution in NOMA-MEC	To minimize the total energy consumption	Cyclic branch network and ResNet	Supervised learning	Channel state information as input; time scheduling and task partitioning as output	Traditional single-connectivity-aided computation offloading scheme, SCA, and NOMAD-based methods
(Qian et al. 2020)	Computation offloading in NOMA-multi-access MEC	To minimize total energy consumption of the IoT devices subject to required latency limit	DRL-based online algorithm	Reinforcement learning	Channel realization in state space; optimal transmission duration in action space; total energy consumption of IoT devices as reward	DRL for OMA
(Lin et al. 2023)	Joint optimization of the task offloading decision and the subchannel assignment in NOMA-MEC	Maximization of task satisfaction ratio of long-term arrival delay-sensitive tasks	DQN (supported by long short-term memory (LSTM))	Reinforcement learning	Input data, delay requirement, waiting time, channel gain, and queue-related information in state space; task offloading and subchannel assignment decision in action space, verification of task completion within recommended delay in reward space	Resource scheduling for single-cell and multi-cell (Noh et al. 2021; Tang and Wong 2020), hierarchical learning-based resource scheduling for multi-cell (Peng and Shen 2020), and local computing-based method
(Yang et al. 2020)	Joint optimization of the task offloading decisions, computation resource allocation, and caching decisions	Minimization of long-term energy consumption of the mobile devices	SAQ-learning and BLA-based MAQ-learning	Reinforcement learning	Network energy consumption at a given time in state space; user task offloading, computing speed allocation from the access point (AP) to users, and cache decision of access point in action space; minimized energy consumption in reward space	SAQ: compared against all local computing, all offloading computing, and non-cache computing. BLA-MAQ: compared against MAQ

Table 2.4 Application of ML for SIC and beamforming

References	Application domain	Objective	ML model(s)	ML technique(s)	Dataset	Compared against
(Xu et al. 2023)	SIC and beamforming for multi-cell NOMA networks	System sum-rate improvement and information overhead reduction	AutoGNN	Unsupervised learning	Data channels and interference channels	ADMM-based approaches
(Sim et al. 2020)	SIC for downlink NOMA networks	Minimization of MSE between transmitted and decoded signals	CNN	Supervised learning	Received superimposed signal as input; decoded user signal as output	Conventional SIC method, perfect SIC decoding (Ding et al. 2015a) and DL-based NOMA system (Kang et al. 2019)
(Awan et al. 2018)	Detector design for uplink NOMA	Solving issues related to high cluster size and increased SIC errors	PLAF	Supervised learning	Received symbols as training data; modulated transmitted symbols as label	MMSE-SIC receiver and NLAF
(Fredj et al. 2022)	Beamforming for uplink cell-free NOMA networks	Network's sum-rate (in Mbps) improvement	D4PG	Reinforcement learning	User SINR in state space; beamforming in action space; sum-rate (in Mbps/Hz) in reward space	MMSE and the simplified conjugate symmetric methods
(Elsayed et al. 2020)	Joint user-cell association and number of beams selection in downlink NOMA networks		Transfer Q-learning	Transfer learning-based reinforcement learning	Average SINR of gNB in state space; user-cell association and number of beams selection in action space; improved average SINR in reward space	Q-learning and BSDC

efficiency but introduces high SIC errors. The proposed design implements a partially linear adaptive filter (PLAF) that takes the received symbols in its input and provides modulated transmitted symbols as output. Reinforcement learning-based techniques are also used for the SIC scheme and beamforming design. For instance, in Fredj et al. (2022), the authors propose a distributed distributional deterministic policy gradient (D4PG) algorithm for beamforming design in the uplink NOMA system for the improvement of its sum-rate. The proposed technique considers user SINR in state space, beamforming matrix in action space, and sum-rate of all users in reward space. In Elsayed et al. (2020), the authors utilize transfer Q-learning for the NOMA network system for joint user-cell association and number of beams selection. The ML architecture assigns the average SINR of next-generation NodeB (gNB) in state space, user-cell association, and the number of beams selection in action space and computes improved average SINR in the reward space. Table 2.4 demonstrates the use of ML-based techniques for SIC and beamforming in NOMA networks.

2.5 Conclusion

This chapter begins with the discussions on the activation functions used in ML for network systems designs and several learning techniques. Besides, it reviews the application of ML-based techniques for NOMA network systems, including user clustering, resource allocation, task offloading, SIC, and beamforming designs.

References

Ahsan W, Yi W, Qin Z, Liu Y, Nallanathan A (2021) Resource allocation in uplink NOMA-IoT networks: A reinforcement-learning approach. IEEE Transactions on Wireless Communications 20(8):5083–5098

Ali Z, Sidhu GAS, Gao F, Jiang J, Wang X (2021) Deep learning based power optimizing for NOMA based relay aided D2D transmissions. IEEE Transactions on Cognitive Communications and Networking 7(3):917–928

Awan DA, Cavalcante RL, Yukawa M, Stanczak S (2018) Detection for 5G-NOMA: An online adaptive machine learning approach. In: 2018 IEEE International Conference on Communications (ICC), IEEE, pp 1–6

Brik B, Ksentini A, Bouaziz M (2020) Federated learning for UAVs-enabled wireless networks: Use cases, challenges, and open problems. IEEE Access 8:53841–53849

Chen W, Zhao S, Zhang R, Chen H, Yang L (2020) Generalized user grouping in NOMA: an overlapping perspective. IEEE Transactions on Wireless Communications 20(5):2876–2887

Ding Z, Adachi F, Poor HV (2015a) The application of MIMO to non-orthogonal multiple access. IEEE transactions on wireless communications 15(1):537–552

Ding Z, Fan P, Poor HV (2015b) Impact of user pairing on 5G nonorthogonal multiple-access downlink transmissions. IEEE Transactions on Vehicular Technology 65(8):6010–6023

Elsayed M, Erol-Kantarci M, Yanikomeroglu H (2020) Transfer reinforcement learning for 5G new radio mmWave networks. IEEE Transactions on Wireless Communications 20(5):2838–2849

Fredj F, Al-Eryani Y, Maghsudi S, Akrout M, Hossain E (2022) Distributed beamforming techniques for cell-free wireless networks using deep reinforcement learning. IEEE Transactions on Cognitive Communications and Networking 8(2):1186–1201

He C, Hu Y, Chen Y, Zeng B (2019) Joint power allocation and channel assignment for NOMA with deep reinforcement learning. IEEE Journal on Selected Areas in Communications 37(10):2200–2210

Higuchi K, Benjebbour A (2015) Non-orthogonal multiple access (NOMA) with successive interference cancellation for future radio access. IEICE Transactions on Communications 98(3):403–414

Kang JM, Kim IM, Chun CJ (2019) Deep learning-based MIMO-NOMA with imperfect SIC decoding. IEEE Systems Journal 14(3):3414–3417

Khan LU, Saad W, Han Z, Hossain E, Hong CS (2021) Federated learning for Internet of Things: Recent advances, taxonomy, and open challenges. IEEE Communications Surveys & Tutorials 23(3):1759–1799

Kim W, Ahn Y, Shim B (2020) Deep neural network-based active user detection for grant-free NOMA systems. IEEE Transactions on Communications 68(4):2143–2155

Kumaresan SP, Tan CK, Ng YH (2020) Efficient user clustering using a low-complexity artificial neural network (ANN) for 5G NOMA systems. IEEE Access 8:179307–179316

Kumaresan SP, Tan CK, Ng YH (2021a) Deep neural network (DNN) for efficient user clustering and power allocation in downlink non-orthogonal multiple access (NOMA) 5G networks. Symmetry 13(8):1507

Kumaresan SP, Tan CK, Ng YH (2021b) Extreme learning machine (ELM) for fast user clustering in downlink non-orthogonal multiple access (NOMA) 5G networks. IEEE Access 9:130884–130894

Lei L, Yuan D, Ho CK, Sun S (2015) Joint optimization of power and channel allocation with non-orthogonal multiple access for 5G cellular systems. In: 2015 IEEE Global Communications Conference (GLOBECOM), IEEE, pp 1–6

Li C, Wang H, Song R (2020) Intelligent offloading for NOMA-assisted MEC via dual connectivity. IEEE Internet of Things Journal 8(4):2802–2813

Li R, Zhao Z, Chen X, Palicot J, Zhang H (2014) TACT: A transfer actor-critic learning framework for energy saving in cellular radio access networks. IEEE Transactions on Wireless Communications 13(4):2000–2011

Lin B, Lai Q, Ghassemlooy Z, Tang X (2021) A machine learning based signal demodulator in NOMA-VLC. Journal of Lightwave Technology 39(10):3081–3087

Lin L, Zhou W, Yang Z, Liu J (2023) Deep reinforcement learning-based task scheduling and resource allocation for NOMA-MEC in Industrial Internet of Things. Peer-to-Peer Networking and Applications 16(1):170–188

Liu P, Li Y, Cheng W, Zhang W, Gao X (2019) Energy-efficient power allocation for millimeter wave beamspace MIMO-NOMA systems. IEEE Access 7:114582–114592

Liu Y, Zhang S, Mu X, Ding Z, Schober R, Al-Dhahir N, Hossain E, Shen X (2022) Evolution of NOMA toward next generation multiple access (NGMA) for 6G. IEEE Journal on Selected Areas in Communications 40(4):1037–1071

Liu YF, Dai YH (2013) On the complexity of joint subcarrier and power allocation for multi-user OFDMA systems. IEEE Transactions on Signal Processing 62(3):583–596

Mahmood MR, Matin MA, Sarigiannidis P, Goudos SK (2022) A comprehensive review on artificial intelligence/machine learning algorithms for empowering the future IoT toward 6G era. IEEE Access 10:87535–87562

Naouri A, Wu H, Nouri NA, Dhelim S, Ning H (2021) A novel framework for mobile-edge computing by optimizing task offloading. IEEE Internet of Things Journal 8(16):13065–13076

Niu S, Liu Y, Wang J, Song H (2021) A decade survey of transfer learning (2010–2020). IEEE Transactions on Artificial Intelligence 1(2):151–166

Noh W, Cho S, et al. (2021) Delay minimization for NOMA-enabled mobile edge computing in industrial Internet of Things. IEEE Transactions on Industrial Informatics 18(10):7321–7331

Pan SJ, Yang Q (2010) A survey on transfer learning. IEEE Transactions on Knowledge and Data Engineering 22(10):1345–1359

Parhi R, Nowak RD (2020) The role of neural network activation functions. IEEE Signal Processing Letters 27:1779–1783

Peng H, Shen X (2020) Deep reinforcement learning based resource management for multi-access edge computing in vehicular networks. IEEE Transactions on Network Science and Engineering 7(4):2416–2428

Qian L, Wu Y, Jiang F, Yu N, Lu W, Lin B (2020) NOMA assisted multi-task multi-access mobile edge computing via deep reinforcement learning for industrial Internet of Things. IEEE Transactions on Industrial Informatics 17(8):5688–5698

Rajasekaran AS, Yanikomeroglu H (2023) Neural network aided user clustering in mmWave-NOMA systems with user decoding capability constraints. IEEE Access 11:45672–45687, DOI 10.1109/ACCESS.2023.3274556

Sim I, Sun YG, Lee D, Kim SH, Lee J, Kim JH, Shin Y, Kim JY (2020) Deep learning based successive interference cancellation scheme in nonorthogonal multiple access downlink network. Energies 13(23):6237

Sun Y, Wang Y, Jiao J, Wu S, Zhang Q (2019) Deep learning-based long-term power allocation scheme for NOMA downlink system in S-IoT. IEEE Access 7:86288–86296

Tang M, Wong VW (2020) Deep reinforcement learning for task offloading in mobile edge computing systems. IEEE Transactions on Mobile Computing 21(6):1985–1997

Van Otterlo M, Wiering M (2012) Reinforcement learning and Markov decision processes. In: Reinforcement learning: State-of-the-art, Springer, pp 3–42

Wang S, Chen M, Liu X, Yin C, Cui S, Poor HV (2020) A machine learning approach for task and resource allocation in mobile-edge computing-based networks. IEEE Internet of Things Journal 8(3):1358–1372

Xiao L, Li Y, Dai C, Dai H, Poor HV (2017) Reinforcement learning-based NOMA power allocation in the presence of smart jamming. IEEE Transactions on Vehicular Technology 67(4):3377–3389

Xu X, Liu Y, Mu X, Chen Q, Jiang H, Ding Z (2023) Artificial intelligence enabled NOMA toward next generation multiple access. IEEE Wireless Communications 30(1):86–94

Yang N, Zhang H, Long K, Hsieh HY, Liu J (2019) Deep neural network for resource management in NOMA networks. IEEE Transactions on Vehicular Technology 69(1):876–886

Yang Z, Liu Y, Chen Y, Al-Dhahir N (2020) Cache-aided NOMA mobile edge computing: A reinforcement learning approach. IEEE Transactions on Wireless Communications 19(10):6899–6915

Yang Z, Chen M, Saad W, Xu W, Shikh-Bahaei M, Poor HV, Cui S (2021) Energy-efficient wireless communications with distributed reconfigurable intelligent surfaces. IEEE Transactions on Wireless Communications 21(1):665–679

Yu H, Fei Z, Zheng Z, Ye N, Han Z (2022) Deep learning-based user activity detection and channel estimation in grant-free NOMA. IEEE transactions on wireless communications 22(4):2202–2214

Yuan Y, Zheng G, Wong KK, Ottersten B, Luo ZQ (2020) Transfer learning and meta learning-based fast downlink beamforming adaptation. IEEE Transactions on Wireless Communications 20(3):1742–1755

Zhai Q, Bolić M, Li Y, Cheng W, Liu C (2021) A Q-learning-based resource allocation for downlink non-orthogonal multiple access systems considering QoS. IEEE Access 9:72702–72711

Zhang H, Zhang H, Long K, Karagiannidis GK (2020) Deep learning based radio resource management in NOMA networks: User association, subchannel and power allocation. IEEE Transactions on Network Science and Engineering 7(4):2406–2415

Zhong R, Liu X, Liu Y, Chen Y, Han Z (2022a) Mobile reconfigurable intelligent surfaces for NOMA networks: Federated learning approaches. IEEE Transactions on Wireless Communications 21(11):10020–10034

Zhong R, Liu X, Liu Y, Chen Y, Wang X (2022b) Path design and resource management for NOMA enhanced indoor intelligent robots. IEEE transactions on wireless communications 21(10):8007–8021

Zhu H, Wu Q, Wu XJ, Fan Q, Fan P, Wang J (2021) Decentralized power allocation for MIMO-NOMA vehicular edge computing based on deep reinforcement learning. IEEE Internet of Things Journal 9(14):12770–12782

Zhu J, Wang J, Huang Y, He S, You X, Yang L (2017) On optimal power allocation for downlink non-orthogonal multiple access systems. IEEE Journal on Selected Areas in Communications 35(12):2744–2757

Resource Allocation (RA) in NOMA-Based Wireless Networks 3

3.1 Introduction

The primary goal of 6G-based next-generation wireless networks (NGWNs) is to enable seamless connectivity and at any time by providing unlimited access to all types of applications. To meet the rapidly growing demands for connected devices and services, these networks aim to provide unrestricted access and uninterrupted connectivity. As the number of users and devices continues to grow, next-generation multiple access (NGMA) systems must efficiently accommodate diverse user requirements. Achieving these goals necessitates the development of innovative resource allocation strategies, enhanced network capacity, and improved EE to optimize overall network performance.

3.2 Resource Allocation Strategies in NOMA

There are several strategies for achieving optimal solutions to the resource allocation problems. Approaches such as exhaustive search, semidefinite relaxation, branch-and-bound, etc. are adopted for resource allocation in NOMA networks. For instance, exhaustive search is a common choice for user pairing and power allocation problems (Islam et al. 2018). In Zakeri et al. (2021), the authors implement semidefinite relaxation-based solution for SIC ordering, subcarrier frequency allocation, and beamforming design problems. The subcarrier frequency allocation problem in Liu et al. (2021) is performed by employing branch-and-bound-based methods.

These approaches provide optimal solutions; however they are computationally expensive. Hence, suboptimal solutions are studied for resource allocation problems in NOMA networks to obtain acceptable performance-computational complexity tradeoffs. For

M. Abdul Matin, M. R. Mahmood, *Machine Learning in Next Generation Multiple Access (NGMA)*, Synthesis Lectures on Communications,
https://doi.org/10.1007/978-3-032-19420-6_3

example, the successive convex approximation method is studied in Liu et al. (2021); Ni et al. (2021) for power allocation problems NOMA networks. In Yu et al. (2022), block coordinate descent, along with successive convex approximation and Dinkelbach methods, is studied for beamforming designs (passive and analog beamforming) and power allocation for intelligent surface-based NOMA.

The aforementioned methods require learning explicit analytical mathematical formulations to provide required solutions. However, with the expansion of network sizes, the complexity of such methods increases and thus restrains their implementation in large-scale networks. Therefore, newer methods/solutions are required to address these issues. With this view, this chapter studies ML-based solutions for resource allocation problems.

3.3 Resource Allocation Parameters in NOMA

From NOMA perspective, network parameters such as power, code, time, frequency/sub-carrier frequency, beamforming, etc. must be assigned to provide quality services to the network users. This section provides an overview on these parameters and discusses how their allocation to the users can affect the network service quality from the user perspective.

- *Power:* In NOMA, power allocation is crucial not only to assist SIC for user signal isolation but also to satisfy QoS constraints of the users. In other words, optimum fairness to all the users in terms of achievable rates, sum-rates, and EE of the NOMA networks can also be ensured by carefully allocating power to the users (Zhu et al. 2017).
- *Code:* Contrary to the PD-NOMA, the CD-NOMA utilizes spreading code for differentiating user data. The orthogonality in CD-NOMA lies in code assignment to the network users, enabling them to provide all the power, time, and frequency resources. Such resource assignment helps in providing network connectivity to the IoT devices with limited computing resources (Jafarkhani et al. 2024).
- *Time/subcarrier frequency:* Time/subcarrier frequency allocation in NOMA systems serves users in terms of latency and throughput. In mobile edge computing-enabled NOMA (MEC-NOMA), time allocation can help the user devices to perform flexible task offloading while fulfilling their latency requirements (Ding et al. 2019). Optimum subcarrier frequency allocation can enhance the network reliability and efficient bandwidth usability. The joint time/sub-frequency and power allocation optimization enhances EE with minimum network energy consumption (Pivoto et al. 2025).
- *SIC ordering and user clustering:* As mentioned in Sect. 1.4, SIC ordering helps in minimizing error propagation while decoding users' signals in NOMA, improving network performance. In case of multiple users present in NOMA networks, user grouping assists in simplifying SIC and thus minimizing SIC error.
- *Beamforming:* Beamforming improves network energy utilization by directing signal beams toward designated area to mitigate interferences, as mentioned in Sect. 1.4.

Sum-rate improvement along with efficient power utilization is possible to achieve by considering power constraint while designing beamforming schemes (Salem et al. 2016; Sun et al. 2018).

This chapter particularly focuses on power allocation problem for NGMA, particularly NOMA, to maximize sum-rate performances. It presents a case study where a two-user and a multiuser NOMA and their resource allocation problems are defined through their mathematical formulations. Later, it discusses ML models to design power allocation schemes for both the NOMA systems while fulfilling QoS constraint(s).

3.4 Mathematical Formulation of NOMA Systems

3.4.1 Two-User NOMA System

This section presents a downlink NOMA network system model, as presented in Fig. 3.1, that considers a BS (with N_{BS} antenna dedicated for each user) and $K = 2$ single-antenna users $N_k = 1$ in the downlink NOMA system. Signals are transmitted by the BS to the two users by following beamformer-based NOMA schemes, explained in Sect. 4.2.5.1.

$$y_k = \boldsymbol{h}_k \boldsymbol{v}_k \sum_{i=1}^{K} \sqrt{p_i} x_i + n_k \tag{3.1a}$$

$$= \boldsymbol{h}_k \boldsymbol{v}_k \sqrt{p_k} x_k + \boldsymbol{h}_k \boldsymbol{v}_k \sqrt{p_{k'}} x_{k'} + n_k. \tag{3.1b}$$

Here, y_k is the received signal at user k, and x_i, $\boldsymbol{v}_k$, and $p_i = \alpha_i P$ are the transmitted signals, transmit precoding vector, and power allocated to user $i \in K$, respectively. The channel vector and noise at user k are denoted by $\boldsymbol{h}_k$ and n_k (with variance σ^2) respectively.

Given the ordering of the channel gains $|\boldsymbol{h}_1 \boldsymbol{v}_1|^2 \leq |\boldsymbol{h}_2 \boldsymbol{v}_2|^2$, user 1 performs direct signal decoding, treating user 2 signal as interference. User 2 performs SIC to remove user 1 signal, demodulates the recovered symbols, and then performs channel decoding to extract its own data bits.

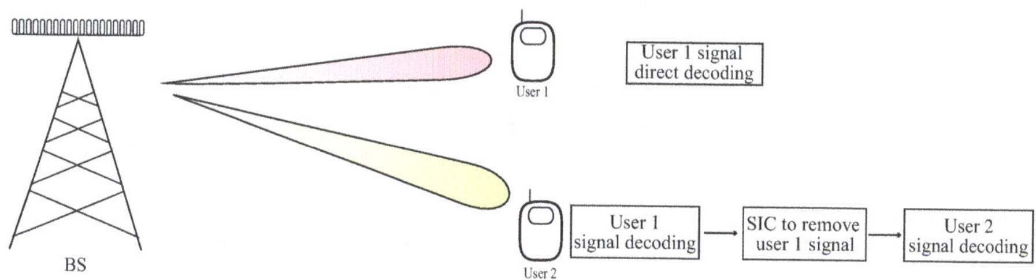

Fig. 3.1 Two-user NOMA system model

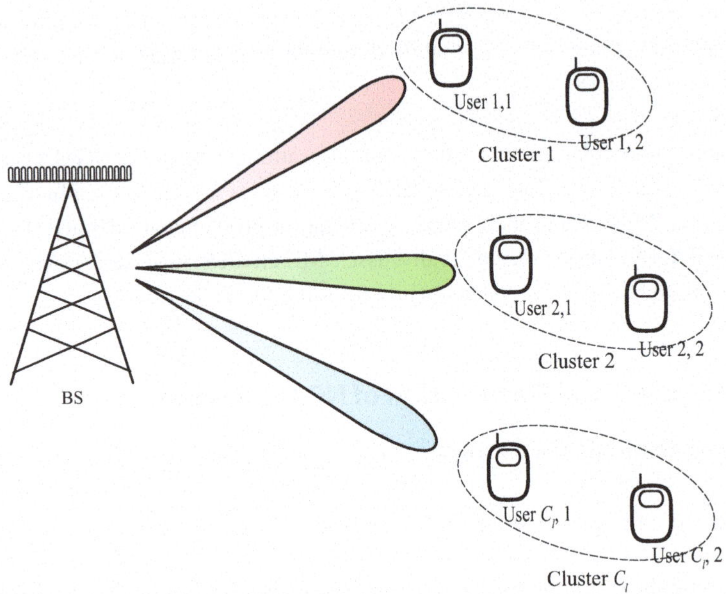

Fig. 3.2 Multiuser NOMA system model

3.4.2 Multiuser NOMA system

This section presents multiuser NOMA system, where a BS (with N_{BS} antenna dedicated for each user) serves multiple users ($K > 2$) simultaneously. This particular model considers user grouping/clustering, where each cluster c_l contains two users ($k \in [1, 2]$). The users are paired on the basis of channel correlation information. Figure 3.2 demonstrates multiuser NOMA system model, which is mathematically represented in (3.2).

$$y_{c_l,k} = \boldsymbol{h}_{c_l,k}\boldsymbol{v}_{c_l}\left(\sum_k \sqrt{p_{c_l,k}}x_{c_l,k}\right) + \sum_{c_l' \neq c_l} \boldsymbol{h}_{c_l,k}\boldsymbol{v}_{c_l'}\left(\sum_k \sqrt{p_{c_l',k}}x_{c_l',k}\right) + n_{c_l,k}. \quad (3.2)$$

Here, $y_{c_l,k}$ is the received signal, $x_{c_l,k}$, $\boldsymbol{h}_{c_l,k}$, and $n_{c_l,k}$ are the transmitted signal, CSI, and noise (with variance σ^2), respectively, for user k in cluster c_l. The beamforming vector for cluster c_l is denoted by v_{c_l}, which is constructed from user with stronger channel gain (Kimy et al. 2013). The inter-cluster interference is assumed to be mitigated due to high correlation between the users in the same cluster (Islam et al. 2018). SIC is performed within the cluster in the same manner as highlighted in Sect. 3.4.1, depending on channel gains. The user with higher channel gain performs SIC to remove the signal of the user with lower channel gains.

3.5 Problem Formulation for Resource Allocation

Two-User NOMA

The achievable/sum-rate expressions for user $k = [1, 2]$ are then given by

$$R_1 = \log_2\left(1 + \frac{p_1|\boldsymbol{h}_1\boldsymbol{v}_1|^2}{p_2|\boldsymbol{h}_1\boldsymbol{v}_1|^2 + \sigma^2}\right) \tag{3.3a}$$

$$R_2 = \log_2\left(1 + \frac{p_2|\boldsymbol{h}_2\boldsymbol{v}_2|^2}{\sigma^2}\right) \tag{3.3b}$$

$$R_{sum} = R_1 + R_2. \tag{3.3c}$$

The system requires to maintain a minimum target rate for user 1 and maximize the sum-rate. The effective sum-rate maximization problem is formulated as the optimum power allocation problem.

$$\max_{\min \alpha_1, \alpha_2} \quad R_{sum} \tag{3.4a}$$

$$\text{s.t.} \quad 0 \leq \{\alpha_1, \alpha_2\} \leq 1 = 1, \tag{3.4b}$$

$$\alpha_1 + \alpha_2 = 1, \tag{3.4c}$$

$$R_1 \geq R_{min}. \tag{3.4d}$$

Multiuser NOMA

The achievable/sum-rate expressions of the multiuser NOMA are given in terms of cluster-based configuration.

$$R_{c_l,1} = \log_2\left(1 + \frac{p_1|\boldsymbol{h}_{c_l,1}\boldsymbol{v}_{c_l}|^2}{p_2|\boldsymbol{h}_1\boldsymbol{v}_1|^2 + \sigma^2}\right) \tag{3.5a}$$

$$R_{c_l,2} = \log_2\left(1 + \frac{p_2|\boldsymbol{h}_{c_l,2}\boldsymbol{v}_{c_l}|^2}{\sigma^2}\right) \tag{3.5b}$$

$$R_{mu-NOMA,sum} = \sum_{c_l}\left(R_{c_l,1} + R_{c_l,2}\right). \tag{3.5c}$$

In this case, the effective sum-rate maximization problem is formulated as optimum power allocation problem, which considers minimum achievable rate for all the users.

$$\max_{\min \alpha_{c_l,1}, \alpha_{c_l,2}} \quad R_{mu-NOMA,sum} \tag{3.6a}$$

$$\text{s.t.} \quad 0 \leq \{\alpha_{c_l,1}, \alpha_{c_l,2}\} \leq 1 = 1, \tag{3.6b}$$

$$\sum_{c_l} (\alpha_{c_l,1} + \alpha_{c_l,2}) = 1, \tag{3.6c}$$

$$R_{c_l,k} \geq R_{min}. \tag{3.6d}$$

3.6 Machine Learning Methodologies for RA

The following discusses a supervised learning model named multilayer perceptron (MLP), an unsupervised learning model, and a deep reinforcement learning model named deep deterministic policy gradient (DDPG) for optimal power allocation with achievable rate constraint.

Supervised Learning Model

Similar to the supervised learning model discussed in Sect. 2.2.1, the MLP model takes channel vectors as input and provides optimal power coefficients as output. In particular, data X_{train} and output Y_{train} consist of the channel vectors $[\boldsymbol{h}_{1,real}, \boldsymbol{h}_{1,imag}, \boldsymbol{h}_{2,real}, \boldsymbol{h}_{2,imag}]^\top$ (fed into the input layer with $4 \times N_{BS}$ neurons) and $[\alpha_1, \alpha_2]^\top$ (fed into output layer with K neurons), respectively. A two-hidden-layered fully connected neural network architecture (with 64 and 32 neurons in consecutive hidden layers) is implemented, where the neurons of the first hidden layer are employed with ReLU activation function and the second hidden layer are employed with sigmoid activation function. Adam optimizer (Kingma 2014) is chosen for updating the weights of the MLP with learning rate of l_r, and MSE is chosen as loss function defined as follows:

$$MSE = \frac{\sum_{j=1}^{J} (Y_{predict} - Y_{train})^2}{J}, \tag{3.7}$$

where J is the data size. Algorithm 1 summarizes the overall learning mechanism.

Algorithm 1: Training of MLP-based supervised learning algorithm

Data: Input data $X_{train} = [\boldsymbol{h}_{1,real}, \boldsymbol{h}_{1,imag}, \boldsymbol{h}_{2,real}, \boldsymbol{h}_{2,imag}]^\top$, learning rate
Output: $Y_{train} = \alpha_1,\ \alpha_2$

```
for each epoch do
    Predict Y_predict = α1, α2 Compute the MSE
    Update the ML parameters using Adam optimizers
end
return Trained neural network;
```

Unsupervised Learning Model

The neural network architecture of the unsupervised learning model is similar to that of the MLP. However, the difference lies in its learning mechanism, i.e., it does not require labeled data, which are the power allocation coefficients. Besides, it employs softmax activation function at the output layer to satisfy constraint (3.4c). The loss function $\mathcal{L}_M$ is defined as follows:

$$\mathcal{L}_M = -(R_1 + R_2) + \mathcal{P} * (R_{min} - R_1), \tag{3.8}$$

where $\mathcal{P}$ is a penalty factor. Algorithm 2 summarizes the overall learning mechanism.

Algorithm 2: Training of unsupervised learning algorithm

Data: Input data $[\boldsymbol{h}_{1,real}, \boldsymbol{h}_{1,imag}, \boldsymbol{h}_{2,real}, \boldsymbol{h}_{2,imag}]^\top$, learning rate
Result: Trained unsupervised learning-based neural network

for *each epoch* **do**
    Obtain α_1 and α_2 Calculate the achievable rates R_1 and R_2 Compute the loss function $\mathcal{L}_M$ from Eq. (3.8)
    Update the ML parameters using Adam optimizers
end
return Trained neural network;

Deep Reinforcement Learning Model

The DDPG algorithm inherits the deep neural network and reinforcement learning for state space with large dimensions. The following defines the state and the action spaces, the reward function, and the neural network architecture.

1. *State space*: The state space contains all the channel information of the users at a definite time-step t, defined as

$$s_t = [\boldsymbol{h}_{R,1,1}, \boldsymbol{h}_{I,1,1}, \boldsymbol{h}_{R,1,2}, \boldsymbol{h}_{I,1,2}, \ldots, \boldsymbol{h}_{R,c_I,1}, \boldsymbol{h}_{I,c_I,1}, \boldsymbol{h}_{R,c_I,2}, \boldsymbol{h}_{I,c_I,2}].$$

2. *Action space*: The action space provides the power allocation coefficients allocated to all the users.

$$\boldsymbol{a}_{c,t} = [\alpha_{1,1}, \alpha_{1,2} \ldots, \alpha_{c_I,1}, \alpha_{c_I,2}].$$

3. *Reward*: The reward function contains the sum-rate and a penalty factor, defined as

$$\mathcal{R}_w = R_{mu-NOMA,sum} - \underbrace{\sum_{c_I}\sum_{k}(R_{min} - R_{c_I,k})}_{\text{Penalty factor}}.$$

4. *Neural network*: The actor network $\mathcal{A}_N$ consists of six layers. The first layer (which is the input layer) and the last layer (which is the output layer) consist of $2N_{BS}K$ and K neurons, respectively. The remaining four consecutive layers (from first to last) contain

256, 256, 32, and 32 neurons, with LReLU activation function (with a negative slope value of 0.2) between each of them. The output layer uses softmax activation function to provide output values.

The critic network C_N is divided into two parts. In the first part, the state information (with input size $2N_{BS}K$) and the actor network outputs (with data size K) are processed separately by two distinct two-layered neural networks, where each neural network layer consists of 256 neurons, with LReLU activation function having negative slope of 0.2. In the second part, three dense layers contain 256, 256 (with LReLU having a negative slope value of 0.2), and 1 (with linear activation function) neurons, respectively. The first dense layer takes the concatenated input in the form $[\boldsymbol{s}_t, \boldsymbol{a}_{c,t}]$, and the last layer provides the predicted action-value function, given by $C_N = \sum_t d_f \mathcal{R}_w$ (where $d_f \in [0, 1]$ is a discount factor).

DDPG uses two stages, namely exploration and exploitation stages for training. The exploration stage allows DDPG to observe new state, predict actions, compute reward, and observe next state. This information is stored in replay buffer. The exploitation stage allows updating of critic and actor network parameters on the basis of replay buffer data (Zou et al. 2023). Algorithm 3 summarizes the overall DDPG learning mechanism.

Algorithm 3: Training of unsupervised learning algorithm

Data: Actor network $\mathcal{A}_N$ and critic network C_N, null replay buffer
Result: Action $\boldsymbol{a}_{c_t}$
for *each episode* **do**
  Observe the given state $\boldsymbol{s}_t$
  for *each time-step t* **do**
    Observe the given state s_t
    Predict action $\boldsymbol{a}_{c,t} = \text{clip}(\boldsymbol{a}_{c,t} + \mathcal{N})$, where $\mathcal{N}$ represents noise process
    Obtain reward $\mathcal{R}_w$ and observe the next-state $\boldsymbol{s}_{t+1}$ and store $(\boldsymbol{s}_t, \boldsymbol{a}_{c,t}, \mathcal{R}_w, \boldsymbol{s}_{t+1})$ in a replay buffer
    Compute q-value $q_t = \mathcal{R}_w + d_f C_N(\boldsymbol{s}_{t+1}, \mathcal{A}_N(\boldsymbol{s}_{t+1}))$ after sampling the batch randomly, and update the critic network
    Update the actor network on the basis of the updated critic network
  end
end
return DDPG learnable parameters;

3.7 Numerical Results

This section presents the numerical results for power allocation schemes in the two-user NOMA system. It considers the Rayleigh channel model given by $\boldsymbol{h}_k = \frac{\boldsymbol{f}_k}{\sqrt{d^{\varpi}}}$, where $\boldsymbol{f}_k$ is the complex Rayleigh coefficient, d $(d_k/d_{c_l,k})$ is the distance between BS and user k

Table 3.1 Simulation configurations

Parameters	Values	Parameters	Values
ϖ	4	B_w	1 MHz
N_f	10 dB	R_{min}	1 bps/Hz
Two-user NOMA			
N_{BS}	4	N_k	1
d_1	400 m	d_2	100 m
Multiuser NOMA			
N_{BS}	16	N_k	1
$d_{c_l,k}$	1–500 m	C_l	4

in the two-user/multiuser NOMA networks, and ϖ is the path loss exponent. Maximum ratio transmission (MRT) precoding is chosen to determine $\boldsymbol{v}_k/\boldsymbol{v}_{c_l}$ (Corvaja and Armada 2015). The noise power is given by $-174 + 10\log_{10}(B_w) + N_f$, where B_w and N_f are bandwidth and noise figure, respectively. The minimum achievable rate R_{min} is chosen to be 1 bps/Hz. Table 3.1 provides the configuration of the NOMA networks.

Both the supervised- and unsupervised-based methods for the two-user NOMA network consider 1000 channel realizations, among which they utilize 900 realizations for training and the rest for evaluation. The MLP algorithm uses datasets prepared with the help of exhaustive search algorithm, which determines the optimum power allocation to meet the objectives provided in Eqs. (3.4a)–(3.4d). Figure 3.3 demonstrates the MSE performance of the MLP-based power allocation scheme at the learning rates of $l_r = 0.1,\ 0.01,\ 0.001$

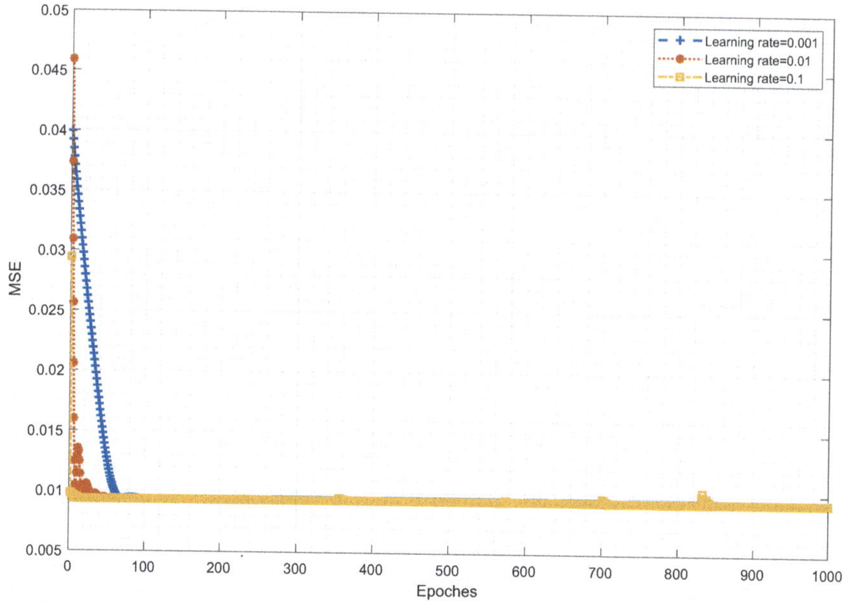

Fig. 3.3 Plot of MSE vs. epochs for MLP-based power allocation for two-user NOMA

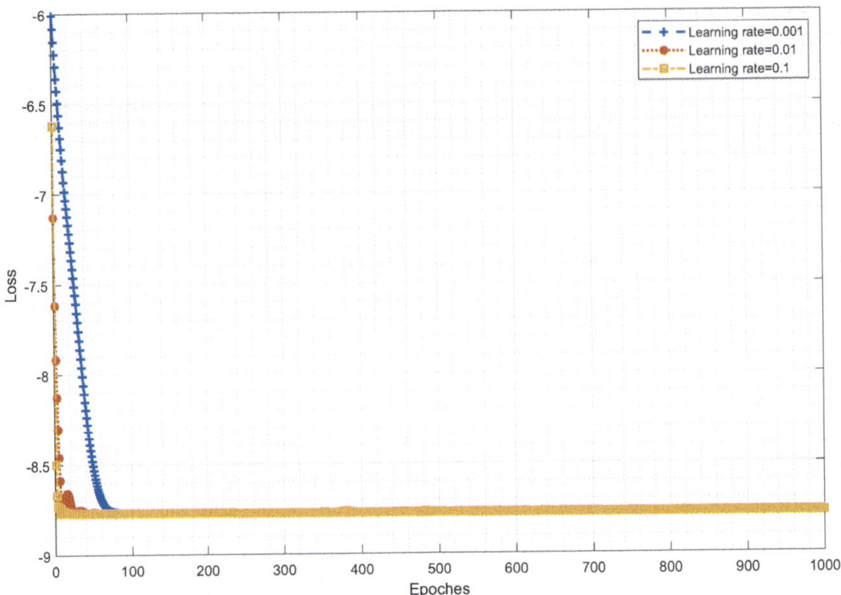

Fig. 3.4 Plot of loss values $\mathcal{L}_M$ vs. epochs for unsupervised machine learning-based power allocation for two-user NOMA

for the transmission power of $P = 30$ dBm. It is observed that the convergence trend for $l_r = 0.001$ is slower compared to that for other l_r values. However, the MSE for all the l_r values converges within 60 epochs. Figure 3.4 demonstrates the loss values ($\mathcal{L}_M$) obtained by the unsupervised machine learning-based power allocation scheme at the same learning rates of $l_r = 0.1,\ 0.01,\ 0.001$ for $P = 30$ dBm. Similar to the observation for MLP-based power allocation, the loss values for all l_r meet at similar points after about 60 epochs. However, minimum loss trend is achieved earlier with $l_r = 0.1$ compared to the other l_r values.

Figure 3.5 demonstrates the achievable rates and the sum-rates in the two-user NOMA network obtained by the exhaustive search, MLP, and unsupervised learning-based power allocation schemes at $l_r = 0.1$ for different values of P. On the one hand, the exhaustive search algorithm is run for all P values with 1000 channel realizations. On the other hand, the ML models are first trained at $P = 30$ dBm with 900 channel realizations. Later, the trained ML models are implemented for different P values over the remaining channel realizations. In case of achievable rates, at 30 dBm, both the exhaustive search and MLP models satisfy the rate constraint for user 1, i.e., R_1 (given in (3.4d)), whereas the unsupervised ML model overshoots R_{min} by about 0.2 bps/Hz. In case of sum-rate performances, the exhaustive search and MLP models outperform unsupervised learning models by about 0.2 bps/Hz.

Figure 3.6 provides the average sum-rate obtained in the multiuser NOMA network by DDPG model for different episodes. For each episode, $t = 10$ time-step is considered. The

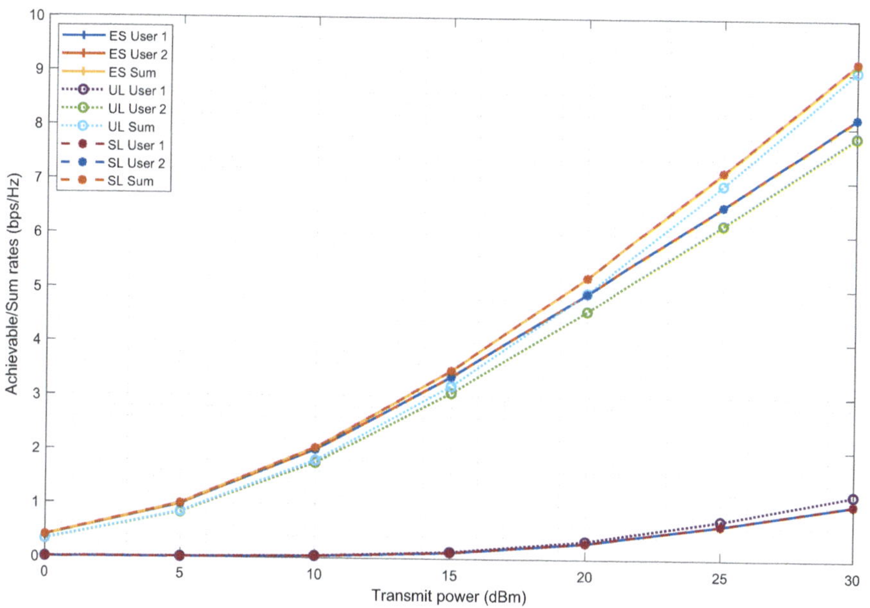

Fig. 3.5 Achievable/sum-rates vs. transmit power (in dBm) for the exhaustive search (ES), MLP, and unsupervised learning (UL)-based power allocation schemes in two-user NOMA

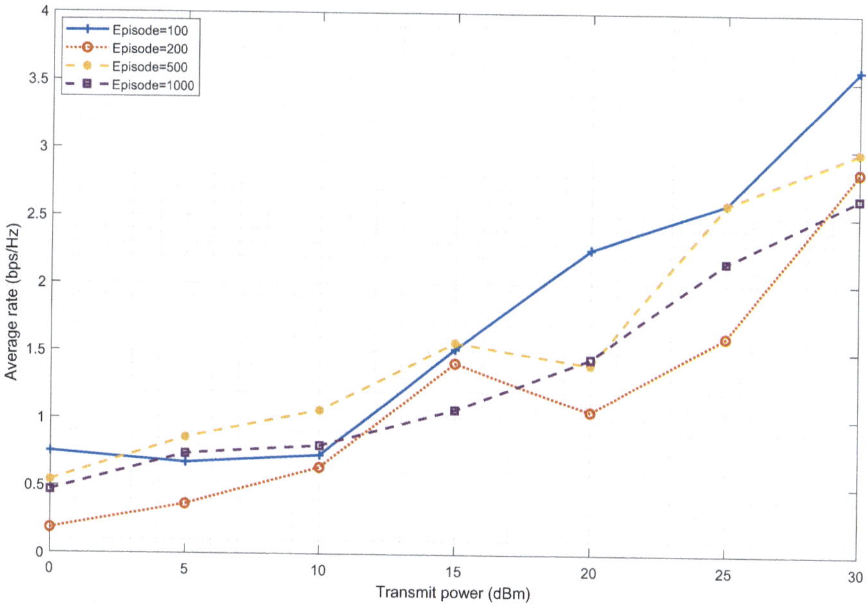

Fig. 3.6 Average sum-rates vs. transmit power (in dBm) for the mu-NOMA with DDPG model-based power allocation

plotted average rate values are obtained by averaging sum-rate values of the last 40% of the total episode data. The average rate for 100 episodes, compared to the other episode values, the performance curve is not stable; whereas, for 1000 episodes, the performance curve is much smoother. It indicates that as the DDPG model keeps learning with larger data and executing its learnings, it keeps providing more stable performances.

3.8 Complexity Analysis of the ML Algorithms

Exhaustive search requires n^2 times operation to determine the optimal power allocation coefficients to both the users. For $n = 100$ power coefficient values for respective users, it goes through 10000 different values. On the contrary, both the trained MLP and the unsupervised ML require $2(I_p L_1 + L_1 L_2 + L_2 K) - (L_1 + L_2 + K) = 6242$ mathematical operations ($I_p = 4N_{BS} = 16$, $L_1 = 64$, $L_2 = 32$, and $K = 2$). Therefore, ML-based methods provide lower complexity compared to the exhaustive search method. It is to note that the supervised learning, in this case, the MLP model, relies on a given dataset. However, the unsupervised learning model relies on a given objective function, which eliminates any dependency on reference data. Therefore, it is advantageous in the scenario in which the reference data are unavailable.

In case of DDPG learning model, if the actor layer contains L_{ac1}, L_{ac2}, L_{ac3}, and L_{ac4} neurons in its hidden layer and K neurons in its output layer, it performs $2(I_p L_{ac1} + L_{ac1} L_{ac2} + L_{ac2} L_{ac3} + L_{ac3} L_{ac4} + L_{ac4} K) - (L_{ac1} + L_{ac2} + L_{ac3} + L_{ac4} + K)$, where $I_p = 2N_{BS}K$. Given $L_{ac1} = L_{ac2} = 256$ and $L_{ac3} = L_{ac4} = 32$, the total number of actor operation is equal to 280504. On the other hand, the critic network performs $2(I_p L_{cr1} + L_{cr1} L_{cr2} + K L_{cr3} + L_{cr3} L_{cr4} + L_{cr2} L_{cr4} L_{cd1} + L_{cd1} L_{cd2} + L_{cd2} L_{cd3}) - (L_{cr1} + L_{cr2} + L_{cr3} + L_{cr4} + L_{cd1} + L_{cd2} + L_{cd3}) = 34081791$ operations when $L_{cr1} + L_{cr1} = L_{cr2} = L_{cr3} = L_{cr4} = L_{cd1} = L_{cd2} = 256$ and $L_{cd3} = 1$. The DDPG-based approach is advantageous when the network requires learning by experience. However, in case the environment parameter changes, such an approach needs to restart the learning process (Zou et al. 2023). Therefore, careful designing of RL-based ML models (in this case, DDPG) is required to obtain satisfactory performance.

3.9 Conclusion

This chapter aims at resource allocation problem in the NOMA network system. It presents two-user and multiuser NOMA system models, followed by problem formulations for resource allocation. It formulated the resource allocation problems as effective sum-rate maximization through optimal power allocation. It presents supervised, unsupervised, and reinforcement learning models for power allocation to meet the required constraints. The numerical results demonstrate the effectiveness of the ML-based approaches for satisfying network system constraints.

References

Corvaja R, Armada AG (2015) Phase noise degradation in massive MIMO downlink with zero-forcing and maximum ratio transmission precoding. IEEE Transactions on Vehicular Technology 65(10):8052–8059

Ding Z, Xu J, Dobre OA, Poor HV (2019) Joint power and time allocation for NOMA–MEC offloading. IEEE Transactions on Vehicular Technology 68(6):6207–6211

Islam SR, Zeng M, Dobre OA, Kwak KS (2018) Resource allocation for downlink NOMA systems: Key techniques and open issues. IEEE Wireless Communications 25(2):40–47

Jafarkhani H, Maleki H, Vaezi M (2024) Modulation and coding for NOMA and RSMA. Proceedings of the IEEE 112(9):1179–1213

Kimy B, Lim S, Kim H, Suh S, Kwun J, Choi S, Lee C, Lee S, Hong D (2013) Non-orthogonal multiple access in a downlink multiuser beamforming system. In: MILCOM 2013-2013 IEEE Military Communications Conference, IEEE, pp 1278–1283

Kingma DP (2014) Adam: A method for stochastic optimization. arXiv preprint arXiv:14126980

Liu L, Sun B, Tan X, Tsang DH (2021) Energy-efficient resource allocation and subchannel assignment for NOMA-enabled multiaccess edge computing. IEEE Systems Journal 16(1):1558–1569

Ni W, Liu X, Liu Y, Tian H, Chen Y (2021) Resource allocation for multi-cell IRS-aided NOMA networks. IEEE Transactions on Wireless Communications 20(7):4253–4268

Pivoto DGS, de Figueiredo FA, Cavdar C, de Lima Tejerina GR, Mendes LL (2025) A comprehensive survey of machine learning applied to resource allocation in wireless communications. IEEE Communications Surveys & Tutorials

Salem FE, Tall A, Altman Z, Gati A (2016) Energy consumption optimization in 5G networks using multilevel beamforming and large scale antenna systems. In: 2016 IEEE Wireless Communications and Networking Conference, IEEE, pp 1–6

Sun X, Yang N, Yan S, Ding Z, Ng DWK, Shen C, Zhong Z (2018) Joint beamforming and power allocation in downlink NOMA multiuser MIMO networks. IEEE Transactions on Wireless Communications 17(8):5367–5381

Yu X, Wang G, Huang X, Wang K, Xu W, Rui Y (2022) Energy efficient resource allocation for uplink RIS-aided millimeter-wave networks with NOMA. IEEE Transactions on Mobile Computing 23(1):423–436

Zakeri A, Khalili A, Javan MR, Mokari N, Jorswieck E (2021) Robust energy-efficient resource management, SIC ordering, and beamforming design for MC MISO-NOMA enabled 6G. IEEE Transactions on Signal Processing 69:2481–2498

Zhu J, Wang J, Huang Y, He S, You X, Yang L (2017) On optimal power allocation for downlink non-orthogonal multiple access systems. IEEE Journal on Selected Areas in Communications 35(12):2744–2757

Zou Y, Liu Y, Mu X, Zhang X, Liu Y, Yuen C (2023) Machine learning in RIS-assisted NOMA IoT networks. IEEE Internet of Things Journal 10(22):19427–19440

Toward Next-Generation NOMA

4

4.1 Introduction

The NGMA is currently undergoing a major transition from an orthogonal to a non-orthogonal paradigm. This shift entails adopting a system in which multiple users or devices share the same bandwidth resources, rather than assigning distinct orthogonal resources to each user or device as in traditional designs. Notably, non-orthogonal transmission schemes have demonstrated superior performance in terms of capacity and spectral efficiency compared to conventional orthogonal approaches. To meet the demanding requirements of 6G networks, researchers are actively developing NOMA-based solutions for NGMA. These efforts aim to enhance the overall efficiency, scalability, and adaptability of NGMA systems through the application of the non-orthogonality principle.

4.2 Variants of NOMA

4.2.1 Space Division Multiple Access (SDMA)

The introduction of multiple antennas in the transmitter and/or receiver allows the exploitation of additional spatial DoF, which allows the network device(s) to access the network systems through utilization of spatial dimension. In space division multiple access (SDMA), the devices are provided with the same time/frequency/code resources, with a different spatial domain (Liu et al. 2022b). In general, linear precoding techniques are used to design transmit/receive beamformers to minimize interference(s). In low/critical network load scenario, SDMA effectively mitigates inter-user interference(s) by utilizing

M. Abdul Matin, M. R. Mahmood, *Machine Learning in Next Generation Multiple Access (NGMA)*, Synthesis Lectures on Communications,
https://doi.org/10.1007/978-3-032-19420-6_4

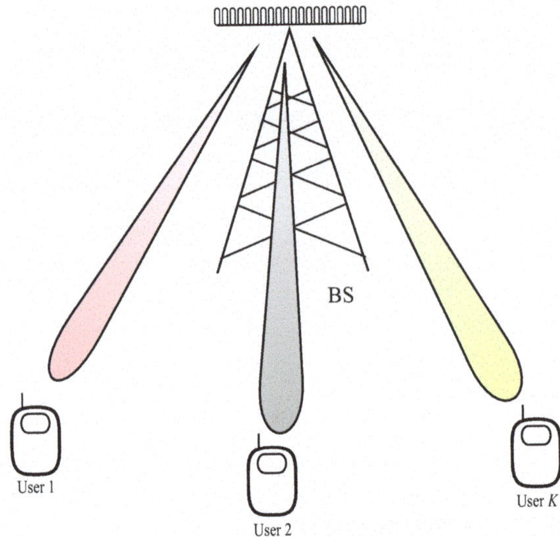

Fig. 4.1 Space division multiple access (SDMA)

spatial DoFs. However, there is a possibility of imperfect interference mitigation if network devices are larger in number compared to available spatial DoFs.

A diagrammatic representation of SDMA-based network system is given in Fig. 4.1. A downlink SDMA consists of a base station (BS) with N_{BS} antennas and user $k \in K$ with a single antenna. Then the received signal by the user k is expressed as follows:

$$y_k = \boldsymbol{h}_k \boldsymbol{v}_k x_k + \boldsymbol{h}_k \sum\nolimits_{i \neq k} v_i x_i + n_k. \tag{4.1}$$

Here, x_k and y_k are the transmitted and received signals, respectively, $\boldsymbol{h}_k$ is the channel between the BS and the user k, and n_k is the noise. The beamforming vector is denoted by $\boldsymbol{v}_{p,k}$, which can be determined by methods such as maximal-ratio combining (MRC), zero-forcing (ZF), MMSE, etc.

The achievable rate is then expressed in the following equation:

$$R_{SDMA,k} = \log_2\left(1 + \underbrace{\frac{|\boldsymbol{v}_k \boldsymbol{h}_k|^2}{\sum_{i \neq k} |\boldsymbol{v}_i \boldsymbol{h}_k|^2 + \sigma^2}}_{\text{SINR}}\right). \tag{4.2}$$

4.2.2 Location Division Multiple Access (LDMA)

Typically, in a far-field scenario, beam steering vectors are utilized to differentiate users in angle domain. However, in a near-field communication scenario, the presence of users in the same/similar angle increases the probability of introducing severe interference, thus negatively affecting simultaneous network accessibility. In this case, the utilization of beam focusing vector allows exploring the location of the users in terms of both distance and angle and providing network access with the help of spherical wave front. This phenomenon introduces a new MA scheme, named location division multiple access (LDMA) (Cui et al. 2022a).

Let us consider a downlink LDMA system with a BS (containing a large N_{BS} antenna array, e.g., uniform linear array (ULA)) and single-antenna users K. Let N_{rf} be the number of radio frequency (RF) chains and $N_{rf} \geq K$. The transmitted signal is denoted by $\boldsymbol{x} = [\sqrt{P\alpha_1}x_1, \sqrt{P\alpha_2}x_2, \ldots, \sqrt{P\alpha_K}x_K]^{\top}$, where P and α_k are total transmitted power and power coefficients for user k, respectively (satisfying $\sum_k \alpha_k = 1$). The following equation expresses the signal received by all the users ($\boldsymbol{y} = [\boldsymbol{y}_1, \boldsymbol{y}_2, \ldots, \boldsymbol{y}_K]$):

$$\boldsymbol{y} = \boldsymbol{H}\boldsymbol{T}\boldsymbol{x} + \boldsymbol{n}. \tag{4.3}$$

Here, $\boldsymbol{H} = [\boldsymbol{h}_1, \boldsymbol{h}_2, \ldots, \boldsymbol{h}_K]$ is the channel matrix between the BS and the users; each of the channel vectors $\boldsymbol{h}_K$ is a function of channel gain and antenna response array, respectively (i.e., $\boldsymbol{h}_K = g_k\boldsymbol{a}_k^r$) (Wu and Dai 2023; Rao et al. 2025). The antenna response array is a function of the BS-user distance and the angle(s) of the signal propagation path. $\boldsymbol{T} = \boldsymbol{T}_{an}\boldsymbol{T}_{dig}$ denotes the hybrid precoding matrix, where $\boldsymbol{T}_{an}$ and $\boldsymbol{T}_{dig}$ are the analog and digital precoder matrix, respectively. In the end, $\boldsymbol{n}$ represents the noise vector with variance σ^2. According to Rao et al. (2025), $\boldsymbol{T}_{an} = \boldsymbol{A}^r = [\boldsymbol{a}_1^r, \boldsymbol{a}_2^r, \ldots, \boldsymbol{a}_K^r]$, and $\boldsymbol{T}_{dig} = \acute{\boldsymbol{H}}^{\mathsf{H}}(\acute{\boldsymbol{H}}\acute{\boldsymbol{H}}^{\mathsf{H}})^{-1}\boldsymbol{\Gamma}$, where $\acute{\boldsymbol{H}} = \boldsymbol{H}\boldsymbol{A}^r$ and $\boldsymbol{\Gamma} = \mathrm{diag}(\lambda_1, \lambda_2, \ldots, \lambda_K)$. To satisfy the equation $||\boldsymbol{T}_{an}\boldsymbol{t}_{dig,k}||^2 = 1$, λ_k is derived as follows:

$$\lambda_k = \frac{1}{\boldsymbol{T}_{an,diag}^{-1}}|g_k|, \tag{4.4}$$

where $\boldsymbol{T}_{an,diag}^{-1} = \mathrm{diag}(\boldsymbol{T}_{an}^{\mathsf{H}}\boldsymbol{T}_{an})^{-1}$. Equation (4.3) is then rewritten as follows:

$$\boldsymbol{y} = \boldsymbol{\Gamma}\boldsymbol{x} + \boldsymbol{n}. \tag{4.5}$$

The sum-rate for LDMA is then given by

$$R_{LDMA} = \sum_k \log_2\left(1 + \frac{P\alpha_k g_k^2}{\boldsymbol{T}_{an,diag}^{-1}\sigma^2}\right). \tag{4.6}$$

4.2.3 Rate-Splitting Multiple Access (RSMA)

Rate-splitting multiple access (RSMA) is based on the principle of the partial decoding of the interference and the partial treatment of that as noise (Mao et al. 2022). For instance,

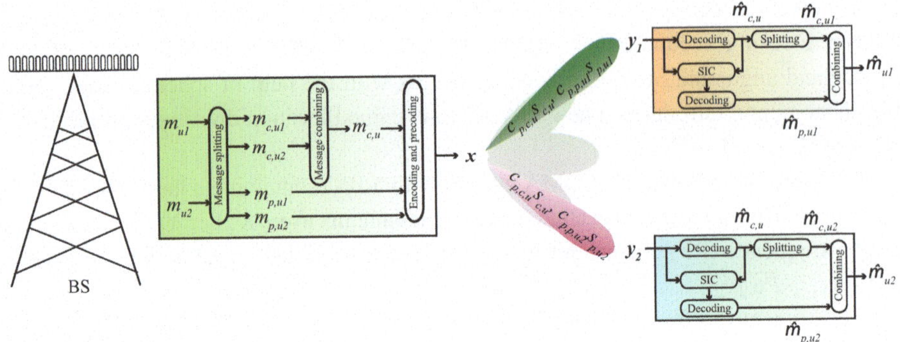

Fig. 4.2 Rate-splitting multiple access (RSMA)

in a two-user ($K = 2$) downlink RSMA-based network system (shown in Fig. 4.2), the message of each user (m_{uk}) is divided into a common part $m_{c,uk}$ and a private part $m_{p,uk}$. The transmit signal is then expressed as $\boldsymbol{x} = \boldsymbol{c}_{p,c,u}s_{c,u} + \sum_k \boldsymbol{c}_{p,p,uk}s_{p,uk}$, where $s_{c,u}$ and $s_{p,uk}$ are encoded as $m_{c,uk}$ and $m_{p,uk}$, respectively. The precoding vectors for the common and the private message parts are denoted by $\boldsymbol{c}_{p,c,u}$ and $\boldsymbol{c}_{p,p,u}$, respectively. The signal received by the user k is written as follows:

$$\begin{aligned} \boldsymbol{y}_k &= \boldsymbol{h}_k^{\mathsf{H}} x + \boldsymbol{n}_k \\ &= \boldsymbol{h}_k^{\mathsf{H}} \boldsymbol{c}_{p,c,u}s_{c,u} + \boldsymbol{h}_k^{\mathsf{H}} \boldsymbol{c}_{p,p,uk}s_{p,uk} + \sum_{i \neq k} \boldsymbol{h}_k^{\mathsf{H}} \boldsymbol{c}_{p,p,ui}s_{p,ui} + \boldsymbol{n}_k. \end{aligned} \quad (4.7)$$

User k decodes $\hat{m}_{c,u}$ from the received signal. On the one hand, it reconstructs $\hat{m}_{c,uk}$ from $\hat{m}_{c,u}$. On the other hand, by employing SIC, it reconstructs the received version of $\hat{m}_{c,u}$ and subtracts from the received signal to obtain $\hat{m}_{p,uk}$. In the end, by combining $\hat{m}_{c,uk}$ and $\hat{m}_{p,uk}$, the user extracts $\hat{m}_{uk}$. The following equations express the achievable rate of user k for decoding $\hat{m}_{c,uk}$ and $\hat{m}_{p,uk}$:

$$R_{RSMA,c,uk} = \log_2(1 + \frac{\boldsymbol{h}_k^{\mathsf{H}} c_{p,c,u}}{\sum_{i \in K} \boldsymbol{h}_k^{\mathsf{H}} c_{p,p,ui} + \sigma_k^2}) \quad (4.8a)$$

$$R_{RSMA,p,uk} = \log_2(1 + \frac{\boldsymbol{h}_k^{\mathsf{H}} c_{p,p,uk}}{\sum_{i \in K, i \neq k} \boldsymbol{h}_k^{\mathsf{H}} c_{p,p,ui} + \sigma_k^2}). \quad (4.8b)$$

4.2.4 Delta-Orthogonal Multiple Access (D-OMA)

The emergence of cell-free networks comes from the necessity of simultaneous network provision to the users in the highly dense wireless network systems. Coordinating a large-

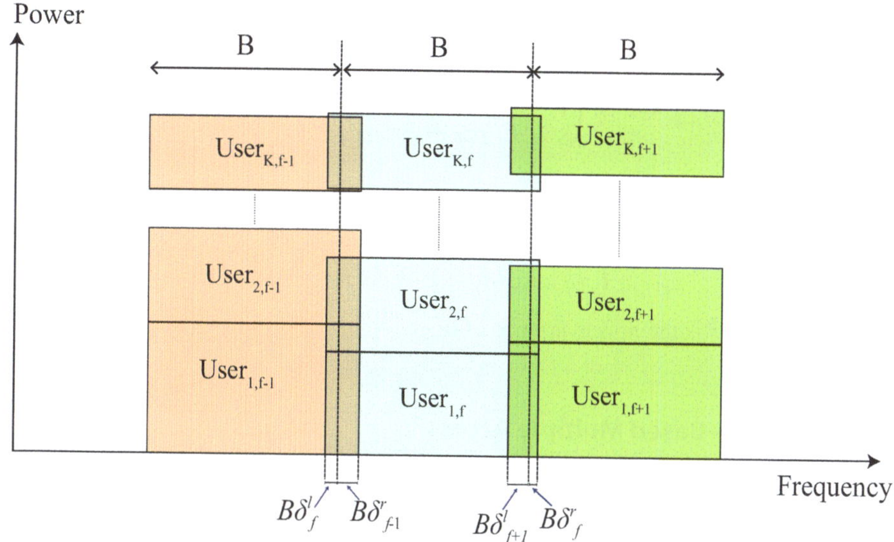

Fig. 4.3 Delta-orthogonal multiple access (D-OMA)

scale deployment of access points (APs)/BSs in the cell-free network systems effectively mitigates high path loss in 6G frequency bands and reduces co-channel interference in overlapping network coverage areas. To support in-band NOMA (where several APs/BSs coordinately transmit signals to multiple users through a sub-band of a given bandwidth) at massive scale, delta-orthogonal multiple access (D-OMA) is proposed in Al-Eryani and Hossain (2019). The NOMA users are clustered with adjacent frequency sub-bands of bandwidth $B = \frac{B_W}{F}$ (B_W is the total bandwidth and F is the number of sub-bands), which are overlapped by a factor δ^l_f, δ^r_f $(0 \leq \delta^l_f, \delta^r_f \leq 1)$, as shown in Fig. 4.3. Given a downlink cell-free network system with S APs, the signal received by user k in sub-band $f \in F$ is

$$
\begin{aligned}
y_k^f = {} & \underbrace{\sum_s h^f_{s,k}\sqrt{p^f_k}\, x^f_k}_{\text{Desired signal}} + \underbrace{\sum_s h^f_{s,k} \sum_{\substack{i \in K \\ i \neq k}} \sqrt{p^f_i}\, x^f_i}_{\text{Intra-cluster interference (ICI)}} \\
& \underbrace{+ \sum_s \sum_{k'=1}^{K_{f-1}} \left(\sqrt{\delta^l_f} + \sqrt{\delta^r_{f-1}}\right) h^f_{s,k'} \sqrt{p^{f-1}_{k'}}\, x^{f-1}_{k'} + \sum_s \sum_{k''=1}^{K_{f+1}} \left(\sqrt{\delta^r_f} + \sqrt{\delta^l_{f+1}}\right) h^f_{s,k''} \sqrt{p^{f+1}_{k'}}\, x^{f+1}_{k}}_{\text{Partial inter-cluster interference (PICI)}} \\
& + \underbrace{\left(\sqrt{1 + \delta^l_f + \delta^r_f}\right) n^f_k}_{\text{Noise}} .
\end{aligned}
\tag{4.9}
$$

The achievable rate of user k at sub-band f is given by

$$R^f_{D-OMA,k} = \log_2\left(1 + \frac{p^f_k \sum_s |h^f_{s,k}|^2}{P_{I,ICI} + P_{I,PICI} + (\hat{\sigma}^f_k)^2}\right). \tag{4.10}$$

Here, $P_{I,ICI} = \sum_{\substack{i\in K\\ i\neq k}} p^f_i \sum_s |h^f_{s,k}|^2$, $P_{I,PICI} = \sum_s \sum_{k'=1}^{K_{f-1}} p^{f-1}_{k'} \left(\sqrt{\delta^l_f} + \sqrt{\delta^r_{f-1}}\right)^2$ $|h^f_{s,k'}|^2 + \sum_s \sum_{k''=1}^{K_{f+1}} p_{k'f+1} \left(\sqrt{\delta^r_f} + \sqrt{\delta^l_{f+1}}\right)^2 |h_{s,k''}f|^2$, and $(\hat{\sigma}^f_k)^2 = \left(1 + \delta^l_f + \delta^r_f\right)$ $(\sigma^f_k)^2$, where $(\sigma^f_k)^2$ is the noise variance of n^f_k.

4.2.5 Antenna-Based Multiple Access

Multi-antenna techniques bring additional degrees of freedom (DoF), which helps in designing effective beamformers for interference mitigation and consequently leading to the improvement of the user's SINR. The following discusses several antenna techniques for NGMA.

4.2.5.1 Beamformer-Based NOMA

Beamformer-based NOMA is influenced by SDMA, where linear beamformers are specifically designed for each user in the network system. In contrast to SDMA, beamformer-based NOMA relies on SIC-based decoding method (Fig. 4.4). Similar to the system configuration mentioned in Sect. 4.2.1, the signal received at user k is given by

$$y_k = \boldsymbol{h}_k \boldsymbol{v}_k x + n_k. \tag{4.11a}$$

$$= \boldsymbol{h}_k \boldsymbol{v}_k \left(\sum_k \sqrt{p_k} x_k\right) + n_k. \tag{4.11b}$$

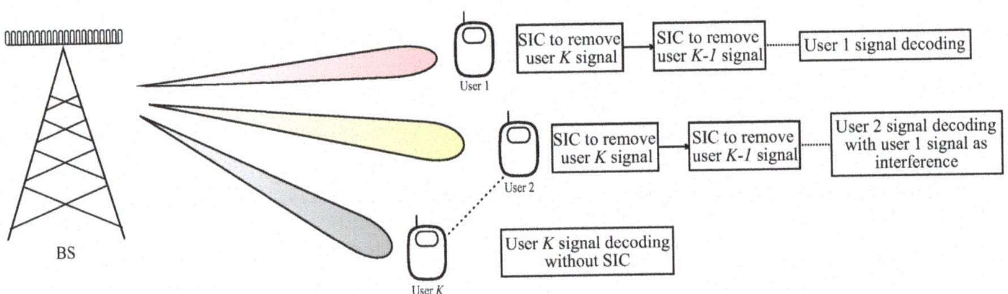

Fig. 4.4 Beamformer-based NOMA

If $\boldsymbol{h}_1\boldsymbol{v}_1 \geq \boldsymbol{h}_2\boldsymbol{v}_2 \geq \ldots \geq \boldsymbol{h}_K\boldsymbol{v}_K$, the user $k = 1$ will remove other signals (i.e., $\sum_{i \neq k} x_i$), remaining with its own signal by means of SIC. Based on this consideration, the SINR expression is given as follows:

$$\gamma_{BF-NOMA,k} = \frac{p_k |\boldsymbol{h}_k \boldsymbol{v}_k|^2}{\sum_{j=k+1}^{K} p_j |\boldsymbol{h}_k \boldsymbol{v}_k|^2 + \sigma^2 {\boldsymbol{v}_k}^{\mathsf{H}} \boldsymbol{v}_k}. \tag{4.12}$$

4.2.5.2 Cluster-Based NOMA

As the network size increases in terms of number of users, the complexity of the determination of optimized decoding order along with optimum beamforming design increases. Cluster-based NOMA aims at minimizing the complexity by grouping/clustering users on the basis of spatial features. The interference among multiple clusters is mitigated by means of beamformers for specified clusters only, reducing the computational load of beamforming design. The intra-cluster interference is minimized with the help of SIC. The following equation represents the signal received by user k ($k \in K$ in every cluster) in cluster c_l (Fig. 4.5):

$$y_{c_l,k} = \boldsymbol{h}_{c_l,k} \boldsymbol{v}_{c_l} \left(\sum_k \sqrt{p_{c_l,k}} x_{c_l,k} \right) + \sum_{c_l' \neq c_l} \boldsymbol{h}_{c_l,k} \boldsymbol{v}_{c_l'} \left(\sum_k \sqrt{p_{c_l',k}} x_{c_l',k} \right) + n_{c_l,k}. \tag{4.13}$$

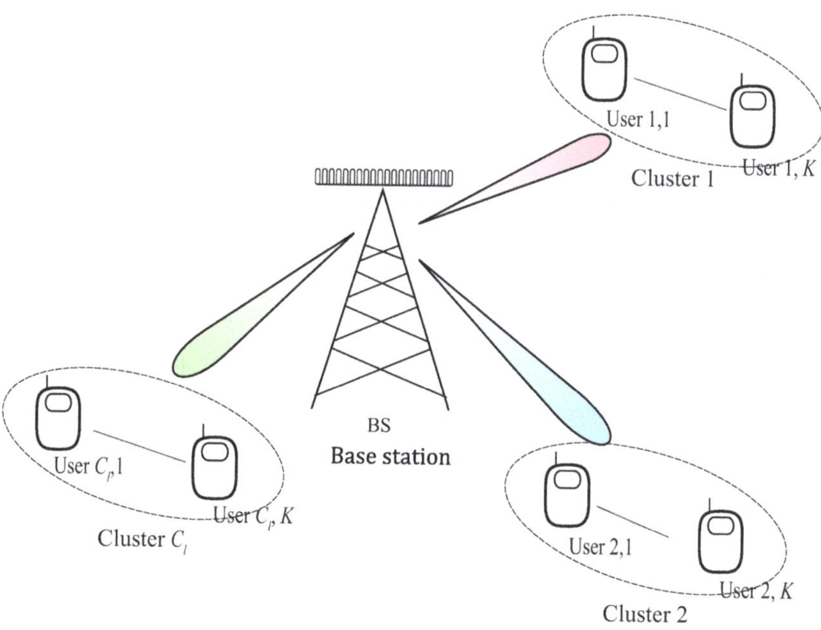

Fig. 4.5 Cluster-based NOMA

The SINR of user k in cluster c_l is given by

$$\gamma_{Cl-NOMA,c_l,k} = \frac{p_{c_l,k}|\boldsymbol{h}_{c_l,k}\boldsymbol{v}_{c_l}|^2}{\underbrace{\sum_{j\neq k}\xi_{k,j}p_{c_l,j}|\boldsymbol{h}_{c_l,k}\boldsymbol{v}_{c_l}|^2}_{\text{Residual ICI}} + \underbrace{\sum_{c_l'\neq c_l}p_{c_l',k}|\boldsymbol{h}_{c_l,k}\boldsymbol{v}_{c_l'}|^2}_{\text{Inter-cluster interference}} + \sigma^2\boldsymbol{v}_{c_l}^{\mathrm{H}}\boldsymbol{v}_{P,c_l}}. \tag{4.14}$$

Here, $\xi \in \{0, 1\}$ is the decoding order of user k, which decides whether one user will perform SIC to decode the other users' signal(s) (e.g., if $\xi_{k,j} = 0$, user k will perform SIC; else, it will directly decode the received signal without performing SIC).

4.2.5.3 Cell-Free NOMA

Cell-free NOMA is realized by multiple APs (coordinated by a CPU), distributed spatially across the network system, serving several users simultaneously over the same time-frequency resources (Ngo et al. 2017). Every user in the network can connect to many APs as possible, which do not restrict the user to a definite cell. Each AP conducts precoding operation on the basis of the channel links among itself and users it serves, which in turn minimizes fronthaul overhead (Le et al. 2021). Given S APs serving C_l clusters, each with K users, the following equation represents signal received by user k in cluster c_l (Fig. 4.6):

$$\begin{aligned} y_{c_l,k} &= \sum_{s}\boldsymbol{h}_{s,c_l,k}\sum_{c_l}\sum_{k}\boldsymbol{v}_{s,c_l}\sqrt{p_{s,c_l,k}}x_{c_l,k} \\ &+ \sum_{s}\sum_{c_l'\in C_L\setminus\{c_l\}}\sum_{k}\sqrt{p_{s,c_l',k}}\boldsymbol{h}_{s,c_l,k}\boldsymbol{v}_{s,c_l'}x_{c_l',k} + n_{c_l,k}. \end{aligned} \tag{4.15}$$

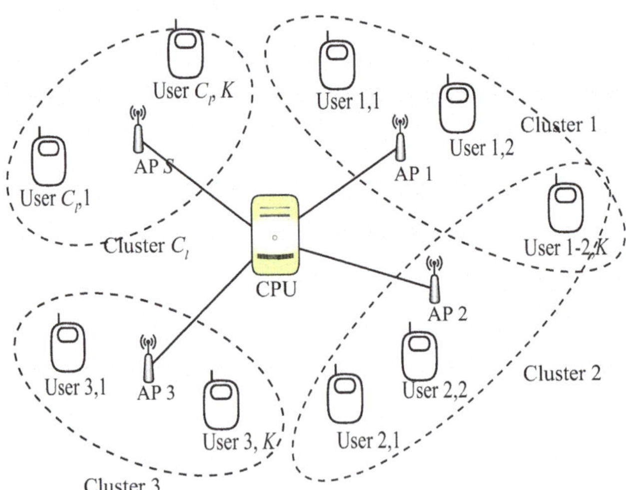

Fig. 4.6 Cell-free NOMA

The SINR of user k in cluster c_l is given by

$$\gamma_{CF-NOMA,c_l,k} = \frac{\left|\mathbb{E}\left\{\sum_s p_{c_l,k}\boldsymbol{h}_{s,c_l,k}\boldsymbol{v}_{s,c_l}\right\}\right|^2}{U_{BF} + ICI_{SIC} + ICI_{R-SIC} + ICI_{k_{c_l}} + \sigma^2 {\boldsymbol{v}_{s,c_l}}^{\mathsf{H}}\boldsymbol{v}_{s,c_l}}, \tag{4.16}$$

where

$$U_{BF} = \mathbb{E}\left\{\left|\left(\sum_s \sqrt{p_{c_l,k}}\boldsymbol{h}_{s,c_l,k}\boldsymbol{v}_{s,c_l} - \mathbb{E}\left\{\sum_s \sqrt{p_{c_l,k}}\boldsymbol{h}_{s,c_l,k}\boldsymbol{v}_{s,c_l}\right\}\right)\right|^2\right\},$$

$$ICI_{SIC} = \sum_{j'=1}^{k-1}\mathbb{E}\left\{\left|\sum_s \sqrt{p_{c_l,j'}}\boldsymbol{h}_{s,c_l,k}\boldsymbol{v}_{s,c_l}\right|^2\right\},$$

$$ICI_{R-SIC} = \sum_{j''=k+1}^{K}\mathbb{E}\left\{\left|\sqrt{\varsigma_k}\sum_s \sqrt{p_{c_l,j''}}\boldsymbol{h}_{s,c_l,k}\boldsymbol{v}_{s,c_l}\right|^2\right\},$$

$$ICI_k = \sum_{c_l'\in C_L\setminus\{c_l\}}\sum_k \mathbb{E}\left\{\left|\sum_s \sqrt{p_{c_l',k}}\boldsymbol{h}_{s,c_l,k}\boldsymbol{v}_{s,c_l'}\right|^2\right\}.$$

Here, U_{BF}, ICI_{SIC}, ICI_{R-SIC}, and ICI_k are beamforming gain uncertainty, intra-cluster interference after SIC, residual interference due to imperfect SIC, and inter-cluster interference, respectively.

4.2.5.4 Fluid-Antenna Multiple Access (FAMA)

Fluid antenna multiple access (FAMA) exploits a fluid antenna architecture, which is defined as any software-oriented controllable structure consisting of fluid, dielectric, or conductive material that adapts itself in terms of its position on the basis of the best available channel (Shah et al. 2024). Figure 4.7 presents an FAMA-enabled network system, where the BS consists of traditional antenna systems, and two users with respective 1D and 2D fluid antenna systems (FASs). On the one hand, the 1D FA consists of N_{an} ports distributed over a length of $W\Lambda$. On the other hand, the 2D FA consists of $MN_{an} = M_{an} \times N_{an}$ ports distributed over an area of $W_1\Lambda \times W_2\Lambda$. The variable(s) W (W_1 and W_2) denote(s) the length of the fluid antennas in 1D (2D) antenna system, and Λ denotes wavelength. As discussed in New et al. (2024), on the basis of aforementioned antenna configuration, several channel models such as simplified, fully correlated, 2D FAS, etc. can be implemented for modeling FAMA network systems. The signal received by user k at the n_pth) antenna port (where $n_p \in N_{an}$ or $n_p \in M_{an}N_{an}$) in the system

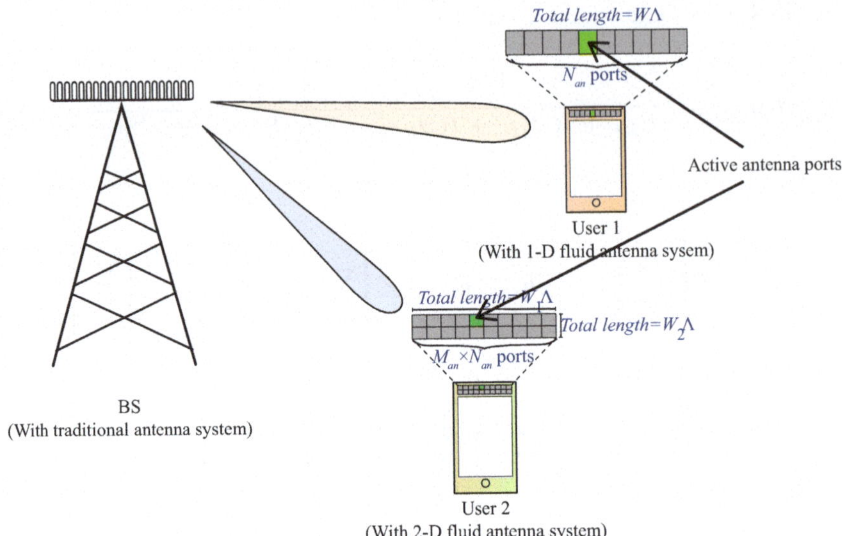

Fig. 4.7 A downlink fluid antenna multiple access (FAMA)-based system, comprising a BS with one traditional antenna system and two users both with FASs

illustrated in Fig. 4.7 is given below:

$$y_k^{n_p} = \boldsymbol{h}_k^{n_p} \boldsymbol{v}_k^{n_p} x_k + \sum_{1 \neq k} \boldsymbol{h}_k^{n_p} \boldsymbol{v}_i^{n_p} x_i + n_k^{n_p}. \tag{4.17}$$

Here, the system model is applicable for two typical scenarios; one is the change of antenna position of user k on the basis of alteration of channel condition, and the other is the change of its antenna position on the symbol-by-symbol basis (Wong et al. 2023, 2022). The former and the later scenarios are termed as slow-FAMA (s-FAMA) and fast-FAMA (f-FAMA), respectively. The achievable rate for user k in both the scenarios is as follows:

$$R_{FAMA} = \begin{cases} \log_2\left(1 + \frac{\left|\boldsymbol{h}_k^{n_p} \boldsymbol{v}_k^{n_p}\right|^2}{\sum_{i \neq k}\left|\boldsymbol{h}_k^{n_p} \boldsymbol{v}_i^{n_p}\right|^2 + \sigma_k^2}\right) & s-\text{FAMA} \\ \\ \log_2\left(1 + \frac{\left|\boldsymbol{h}_k^{n_p} \boldsymbol{v}_k^{n_p} x_k\right|^2}{\sum_{i \neq k}\left|\boldsymbol{h}_k^{n_p} \boldsymbol{v}_i^{n_p} x_i\right|^2 + \sigma_k^2}\right) & f-\text{FAMA}. \end{cases} \tag{4.18}$$

It is to be noted that in s-FAMA, the FAS-enabled device (in this case, the user k device) chooses its antenna port to maximize the received SINR based on channel information. In contrast, in f-FAMA, the user k device considers SINR maximization at each symbol interval. For this reason, the SINR expression for f-FAMA includes transmitted data symbol(s) by the interfering users.

4.3 Enabling Technologies for Next-Generation NOMA

4.3.1 Terahertz (THz) Communication

THz band, with larger offered bandwidth, has the capability of accommodating plenty of users and incorporating several applications. The short wavelength offered by THz necessitates the incorporation of super-large number of antennas in the MIMO-based network systems for exploiting spatial diversity. NOMA is ineffective in a situation where the number of users (with independent channels) is smaller than the number of antennas. This is because in this case, the channels are almost orthogonal. However, the correlation among the low-rank channels of the THz band is high due to the limited-scattering transmission phenomenon. In this case, the integration of NOMA with THz is effective in improving the spectral efficiency in THz-based network systems (Liu et al. 2022a). Though massive connectivity is improved through THz-NOMA, the acquisition of subchannel and power allocation is difficult in high user intensity scenario. This raises the necessity of investigation of resource allocation and their optimization for energy efficient THz-NOMA-based network systems.

4.3.2 Reconfigurable Intelligent Surface (RIS)

The integration of NOMA and reconfigurable intelligent surface (RIS) is explored to provide low latency, massive connectivity, enhanced data rates, and extended network coverage with low power consumption. Traditional wireless network systems often struggle in guaranteeing quasi-degradation due to the fact that the channels are dependent on the propagation environments. The scope of reconfiguring the channels is also not possible. Therefore, RIS-based wireless network allows the required altering of the channels along with the reduction of the transmission power (Zhu et al. 2020). In addition, RIS-NOMA provides greater degree of freedom in the communication scenario where the channel gains of users are identical. RIS is also useful in relieving the constraint of number of antennas at the transmitter/receiver side in NOMA networks by adjusting the phase shifts of RISs.

4.3.3 Integrated Sensing and Communications (ISaC)

Integrated sensing and communications (ISaC) bring the advantages of sensing capabilities and reducing the weights of equipment, which are useful for network scenarios such as V2X and UAV (Liu et al. 2020). In ISaC, sensing and communication signals are different in terms of the waveform they employ. Therefore, if sensing and communication have similar priority, the combination of both the sensing and communication signals in same time/spectrum resource(s) is difficult. If combined in the same time/spectrum resource(s), the simultaneous decoding (of the communication signals) and analysis (of the sensing signals) from the received signal are difficult.

NOMA is better than OMA due to the following reasons (Wang et al. 2022; Cui et al. 2022b): At first, NOMA enables the division of resource block(s) to several users. Besides, the superimposition of communication and sensing signals, each with different power allocation, can be performed in the same time/frequency resource(s). Later, the incorporation of SIC helps in analyzing the received signals to extract the communication and sensing signals.

4.3.4 Visible Light Communication (VLC)

Visible light communication (VLC) provides energy effective communication systems for indoor environments (Zhang et al. 2016). It also utilizes unlicensed spectrum, high confidentiality, and also efficient in terms of implementation. Though in VLC, high transmit SNR and constant channel state are observed, the modulation bandwidth is small. To improve the SE, multiple access techniques must be designed accordingly (Liu et al. 2022b). NOMA is attractive due to the fact that each VLC transmitter serves limited users. Therefore, a small number of SIC operations are required, consequently easing the control of outage probabilities for all SICs. Also, high transmit SNR and slow channel variations enhance throughput and enable low-complexity channel estimation, respectively. Furthermore, VLC-NOMA can exploit transmission angles and field of view to obtain significant channel differences.

4.3.5 Non-terrestrial Networks

Non-terrestrial networks are effective for applications/use-cases such as, disaster management, television broadcasting, navigation, remote sensing, etc. With the evolution of network systems, non-terrestrial networks will support terrestrial network systems for ever-increasing network subscribers. The advantage of wider network coverage by non-terrestrial, its faster deployment along with its easy management of large data streaming and high bandwidth delivery (compared to terrestrial networks) assists in providing connection to network devices in large scale (Ahmed et al. 2024).

NOMA-based non-terrestrial network, such as NOMA-UAV, utilizes its deployment region, which is high-altitude area, providing stronger channel link, compared to terrestrial network scenario. Also, in case of downlink system, it can help in meeting data rate demands through the utilization of their asymmetricity. In case of uplink system, NOMA-UAV can attain enhanced degree of freedom through the achievement of larger macro-diversity than terrestrial users. This enhances the overall system performance by allowing NOMA-UAV to utilize multiple BSs for cooperative SIC or apply precoding to improve throughput or fairness.

4.4 Application/Use Cases of Next-Generation NOMA

4.4.1 Robotic Communications

Robotic communication-based network systems allow robots to exchange information among themselves through APs/BSs, thus mitigating the computational burden and implementation cost. However, robot incorporated network systems suffer from fast-changing channel conditions. In addition, the number of connected robots in robotic communication scenario is large. Many of them differ in terms of their static/mobility state, data size requirements, and delay sensitivity. Such heterogeneity raises the concern of network resource management. By adopting NOMA in such a case, dynamic assignment of static/mobile robot/other devices into same cluster is possible. Effective resource allocation and path planning can improve the spectrum efficiency and security of the robot-incorporated networks (Zhong et al. 2022; Luo et al. 2021).

4.4.2 Virtual/Augmented Reality (VR/AR)

Virtual/augmented reality (VR/AR)-enabled services are sensitive to latency with highly demanding transmit rate and reliability. In this case, traditional OMA schemes fail to meet such criteria. The reason is that network resources in OMA-based schemes are allocated by fixed block sizes. This indicates that the users have to wait for its turn to utilize the resource(s). This introduces delay in service provision. By incorporating NOMA, multiple users can utilize the same network resource for delay-sensitive service(s). Through effective network resource distribution, high spectral efficiency and reliability can be ensured.

4.4.3 Massive Machine-Type Communication (mMTC)

Massive machine-type communication (mMTC) helps in realizing several use cases, such as intelligent transport systems, industry 4.0, and smart cities. The evolution of MTC to mMTC and massive and critical MTC (MC-MTC) brings challenges in meeting 6G QoS requirements. Random access techniques are one of the possible measures to meet such requirements. Compared to grant-based transmission schemes, grant-free transmission scheme improves network connectivity by reducing the signaling overhead. One drawback of grant-free transmission is that it provides limited SE in low channel conditions or in the case of required user data rate being low. NOMA balances the network traffic by adjusting power level allocated to user. Apart from grant-free NOMA, researchers have also explored semi-grant-free NOMA to avoid detection failure (Ding et al. 2019).

4.4.4 Mobile Edge Computing (MEC) Networks

Mobile edge computing (MEC) networks allow the network processing capabilities to be shifted to network edge, which reduces the computation burden of the centralized data center. With the introduction and massive implementation of services such as VR/AR, MEC is a promising solution. With its integration with NOMA, the computational burden on central data center and task processing delay are reduced; also accommodation of a higher number of users along with higher task offloading transmission rates is ensured. Such integration is an attractive enabler for applications such as autonomous vehicles and smart healthcare, which emphasizes the consideration of task processing cost and optimum task offloading service(s) (Liu et al. 2022b; Ahmed et al. 2024).

4.5 Conclusion

This chapter discusses NOMA-based solutions for developing NGMA-based wireless networks. It initiates with discussions on different NOMA variants, namely SDMA, LDMA, RSMA, and D-OMA. The discussion on NOMA variants also includes antenna-based architectures, which employ advanced beamforming techniques. Later, it highlights enabling technologies for next-generation NOMA and several applications/use cases offered by next-generation NOMA.

References

Ahmed A, Xingfu W, Hawbani A, Yuan W, Tabassum H, Liu Y, Qaisar MUF, Ding Z, Al-Dhahir N, Nallanathan A, et al. (2024) Unveiling the potential of NOMA: A journey to next generation multiple access. IEEE Communications Surveys & Tutorials

Al-Eryani Y, Hossain E (2019) The D-OMA method for massive multiple access in 6G: Performance, security, and challenges. IEEE Vehicular Technology Magazine 14(3):92–99

Cui M, Wu Z, Lu Y, Wei X, Dai L (2022a) Near-field MIMO communications for 6G: Fundamentals, challenges, potentials, and future directions. IEEE Communications Magazine 61(1):40–46

Cui Z, Hu J, Cheng J, Li G (2022b) Multi-domain NOMA for ISaC: Utilizing the DoF in the delay-Doppler domain. IEEE Communications Letters 27(2):726–730

Ding Z, Schober R, Fan P, Poor HV (2019) Simple semi-grant-free transmission strategies assisted by non-orthogonal multiple access. IEEE Transactions on Communications 67(6):4464–4478

Le QN, Nguyen VD, Dobre OA, Nguyen NP, Zhao R, Chatzinotas S (2021) Learning-assisted user clustering in cell-free massive MIMO-NOMA networks. IEEE Transactions on Vehicular Technology 70(12):12872–12887

Liu F, Masouros C, Petropulu AP, Griffiths H, Hanzo L (2020) Joint radar and communication design: Applications, state-of-the-art, and the road ahead. IEEE Transactions on Communications 68(6):3834–3862

Liu Y, Yi W, Ding Z, Liu X, Dobre OA, Al-Dhahir N (2022a) Developing NOMA to next generation multiple access: Future vision and research opportunities. IEEE Wireless Communications 29(6):120–127

Liu Y, Zhang S, Mu X, Ding Z, Schober R, Al-Dhahir N, Hossain E, Shen X (2022b) Evolution of NOMA toward next generation multiple access (NGMA) for 6G. IEEE Journal on Selected Areas in Communications 40(4):1037–1071

Luo R, Tian H, Ni W (2021) Communication-aware path design for indoor robots exploiting federated deep reinforcement learning. In: 2021 IEEE 32nd Annual International Symposium on Personal, Indoor and Mobile Radio Communications (PIMRC), IEEE, pp 1197–1202

Mao Y, Dizdar O, Clerckx B, Schober R, Popovski P, Poor HV (2022) Rate-splitting multiple access: Fundamentals, survey, and future research trends. IEEE Communications Surveys & Tutorials 24(4):2073–2126

New WK, Wong KK, Xu H, Wang C, Ghadi FR, Zhang J, Rao J, Murch R, Ramírez-Espinosa P, Morales-Jimenez D, et al. (2024) A tutorial on fluid antenna system for 6G networks: Encompassing communication theory, optimization methods and hardware designs. IEEE Communications Surveys & Tutorials

Ngo HQ, Ashikhmin A, Yang H, Larsson EG, Marzetta TL (2017) Cell-free massive MIMO versus small cells. IEEE transactions on wireless communications 16(3):1834–1850

Rao C, Ding Z, Cumanan K, Dai X (2025) LDMA-NOMA beamforming in near-field communication systems. IEEE Transactions on Vehicular Technology 74(10):15881–15893

Shah AS, Ali Karabulut MA, Cinar E, Rabie KM (2024) A survey on fluid antenna multiple access for 6G: A new multiple access technology that provides great diversity in a small space. IEEE Access 12:88410–88425

Wang Z, Liu Y, Mu X, Ding Z (2022) NOMA inspired interference cancellation for integrated sensing and communication. In: ICC 2022-IEEE International Conference on Communications, IEEE, pp 3154–3159

Wong KK, Tong KF, Chen Y, Zhang Y (2022) Fast fluid antenna multiple access enabling massive connectivity. IEEE Communications Letters 27(2):711–715

Wong KK, Morales-Jimenez D, Tong KF, Chae CB (2023) Slow fluid antenna multiple access. IEEE Transactions on Communications 71(5):2831–2846

Wu Z, Dai L (2023) Multiple access for near-field communications: SDMA or LDMA? IEEE Journal on Selected Areas in Communications 41(6):1918–1935

Zhang X, Gao Q, Gong C, Xu Z (2016) User grouping and power allocation for NOMA visible light communication multi-cell networks. IEEE communications letters 21(4):777–780

Zhong R, Liu X, Liu Y, Chen Y, Wang X (2022) Path design and resource management for NOMA enhanced indoor intelligent robots. IEEE transactions on wireless communications 21(10):8007–8021

Zhu J, Huang Y, Wang J, Navaie K, Ding Z (2020) Power efficient IRS-assisted NOMA. IEEE Transactions on Communications 69(2):900–913

Conclusion and Future Research 5

5.1 Summary

This book focuses on next-generation multiple access (NGMA) empowered by machine learning (ML) techniques to develop next-generation wireless communication systems. Chapter 1 discusses several multiple access (MA) techniques and their standardization. The discussion then continues to non-orthogonal multiple access (NOMA), along with its driving force, prominent NOMA schemes, and issues with its practical implementation. Chapter 2 discusses the fundamentals of ML mechanisms. Besides, it overviews their applications in NOMA-based network systems. Chapter 3 focuses on resource allocation in NOMA network systems. It discusses the transmission and reception model of NOMA networks, problem formulation, and ML techniques for resource allocation along with numerical results. Chapter 4 discusses NOMA-based solutions for NGMA-based wireless networks, including NOMA variants, antenna techniques, and enabling technologies for next-generation NOMA. The chapter also includes several applications/use cases that can be enhanced by adopting NOMA.

5.2 Future Research Directions

This section provides possible future research directions toward the development of NOMA/NGMA-based wireless network systems exploiting machine learning.

- Most of the user grouping/clustering surveyed in Table 2.1 of Chap. 2 considers CSI/path loss for training ML-based methods. An alternative approach to CSI/path loss information-oriented ML training for user clustering can be the utilization of location-

M. Abdul Matin, M. R. Mahmood, *Machine Learning in Next Generation Multiple Access (NGMA)*, Synthesis Lectures on Communications,
https://doi.org/10.1007/978-3-032-19420-6_5

based information for ML training (Rajasekaran and Yanikomeroglu 2023). The near-field communication system can provide the opportunity of utilizing user location due to spherical wave front feature (Rao et al. 2025). Therefore, the combination of LDMA and NOMA holds significant potential in developing high-data rate and beam-efficient enabled near-field communication system. ML methods can also be considered as a viable alternative in this regard.

- Tasks such as user clustering, resource allocation, etc. with multiple network constraints by means of ML methods generally require sufficient computational capability and data availability. However, that is not always the case. ML methods such as meta learning (also called *learning to learn* from experience), federated learning, and transfer learning address the data unavailability and computational capability issues. Therefore, utilizing such approaches for network system optimization is promising a research direction. However, issues such as robustness of the learning model and convergence analysis in such methods must be considered (Liu et al. 2023).
- Integration of several network system enablers provides opportunities to improve network system scenarios and provide effective services to the connected users. For instance, RIS has been incorporated with NOMA-based THz communications, ISaC, VLC, and non-terrestrial networks to solve signal propagation-related issues. This has created several research opportunities for network performance enhancement through optimum power and phase shift control, channel estimation and beamforming, user grouping, signal propagation control, etc. The studies conducted in Xu et al. (2021), Thoong et al. (2024), Lin et al. (2023) have addressed these tasks by employing ML-based methods to achieve satisfactory network performances in terms of sum-rates, EE, and network security. The ML methods provide suboptimal solutions to the defined optimization problems; therefore appropriate ML methods must be chosen for dynamic network scenario (Liu et al. 2024).
- ML-based methods are also used in scenarios such as robotic communications for trajectory designs, VR/AR for image recognition, speech-to-text transcription, relevant web search selection, mMTC for network congestion avoidance and packet delay minimization, and MEC for task and resource allocation problems (Zhong et al. 2022; Zhang et al. 2022; Sharma and Wang 2019; Zhu et al. 2021). However, scalable ML-based methods for dynamic network environment with limited computational capabilities of the network devices are an open challenging problem. Therefore, adaptive learning mechanisms are required to address such issues Luo et al. 2022.

References

Lin Y, Wang K, Ding Z (2023) Unsupervised machine learning-based user clustering in THz-NOMA systems. IEEE Wireless Communications Letters 12(7):1130–1134

Liu R, Guo K, Li X, Dev K, Khowaja SA, Tsiftsis TA, Song H (2024) RIS-empowered satellite-aerial-terrestrial networks with PD-NOMA. IEEE Communications Surveys & Tutorials 26(4):2258–2289

Liu X, Deng Y, Nallanathan A, Bennis M (2023) Federated learning and meta learning: Approaches, applications, and directions. IEEE Communications Surveys & Tutorials 26(1):571–618

Luo R, Ni W, Tian H, Cheng J (2022) Federated deep reinforcement learning for RIS-assisted indoor multi-robot communication systems. IEEE Transactions on Vehicular Technology 71(11):12321–12326

Rajasekaran AS, Yanikomeroglu H (2023) Neural network aided user clustering in mmWave-NOMA systems with user decoding capability constraints. IEEE Access 11:45672–45687, https://doi.org/10.1109/ACCESS.2023.3274556

Rao C, Ding Z, Cumanan K, Dai X (2025) LDMA-NOMA beamforming in near-field communication systems. IEEE Transactions on Vehicular Technology 74(10):15881–15893

Sharma SK, Wang X (2019) Toward massive machine type communications in ultra-dense cellular IoT networks: Current issues and machine learning-assisted solutions. IEEE Communications Surveys & Tutorials 22(1):426–471

Thoong NQ, Cheema AA, Khosravirad SR, Dobre OA, Duong TQ (2024) Channel estimation for reconfigurable intelligent surface-aided 6G NOMA systems using CNN-based quantum LSTM model. In: 2024 IEEE 100th Vehicular Technology Conference (VTC2024-Fall), IEEE, pp 1–5

Xu X, Chen Q, Mu X, Liu Y, Jiang H (2021) Graph-embedded multi-agent learning for smart reconfigurable THz MIMO-NOMA networks. IEEE Journal on Selected Areas in Communications 40(1):259–275

Zhang Z, Wen F, Sun Z, Guo X, He T, Lee C (2022) Artificial intelligence-enabled sensing technologies in the 5G/Internet of Things era: From virtual reality/augmented reality to the digital twin. Advanced Intelligent Systems 4(7):2100228

Zhong R, Liu X, Liu Y, Chen Y, Wang X (2022) Path design and resource management for NOMA enhanced indoor intelligent robots. IEEE transactions on wireless communications 21(10):8007–8021

Zhu H, Wu Q, Wu XJ, Fan Q, Fan P, Wang J (2021) Decentralized power allocation for MIMO-NOMA vehicular edge computing based on deep reinforcement learning. IEEE Internet of Things Journal 9(14):12770–12782

Index

M. Abdul Matin, M. R. Mahmood, *Machine Learning in Next Generation Multiple Access (NGMA)*, Synthesis Lectures on Communications,
https://doi.org/10.1007/978-3-032-19420-6

The manufacturer's authorised representative in the EU is Springer Nature Customer Service Centre GmbH, Europaplatz 3, 69115 Heidelberg, Germany. If you have any concerns regarding our products, please contact ProductSafety@springernature.com

Printed and bound by CPI Group (UK) Ltd, Croydon, CR0 4YY
07/07/2026
02160926-0004